The Nature of the Economy

Ricardo F. Crespo

The Nature of the Economy

Aristotelian Essays on the Philosophy and Epistemology of Economics

palgrave
macmillan

Ricardo F. Crespo
Department of Economics
IAE Business School
Universidad Austral and researcher at
the National Council of Scientific
Research (CONICET, Argentina)
Pilar, Argentina

ISBN 978-3-031-02452-8   ISBN 978-3-031-02453-5 (eBook)
https://doi.org/10.1007/978-3-031-02453-5

This Palgrave Macmillan imprint is published by the registered company Springer Nature Switzerland AG
The registered company address is: Gewerbestrasse 11, 6330 Cham, Switzerland

Preface

There is a story behind every book. The story behind this one goes a long way back. My father held a Ph.D. and was a professional economist who worked at banks his entire life. However, he was also a part-time Economics professor. For some years, he taught "Economic Policy" at Universidad de Mar del Plata on Saturday mornings. We lived in Buenos Aires, 400 km away from Mar del Plata, and, every Friday after work, my father drove our family to Mar del Plata to deliver his classes the following day. We would enjoy the rest of the weekend there—a nice seaside city, though a bit windy, rainy, and cold in the winter (South Atlantic Ocean). I used to sit behind him in the car, and we would talk a lot about economics. Ever since, my ambition has been to ascertain the nature of the economic system, its components, motivations, institutions, and the scope and adequate methods of economics. Hence, after graduating from high school, I studied undergraduate economics. I was fascinated by economic logic. I used to study with a group of friends, and we would discuss the mutual influences of the great set of variables underlying economic events at length; we even role-played as representatives of the Central Bank, the Economics Ministry, the unions, the corporations, and consumers, and we designed economic policies. Once I graduated, I worked in the financial area for seven years.

Unfortunately, I found but a few answers about the nature of the economy during those years, and I remained unsatisfied. I realized that this was a philosophic matter, and thus I decided to study philosophy, starting with an undergraduate program. Then, I focused my philosophy Ph.D. dissertation on philosophy of economics. I understood that my questions were not easy to answer, and that the journey would be long. My further studies, including my Ph.D. dissertation on economics, have delved deeper into the nature and scope of economics, as well as into the meaning of the economy and its components.

My latest book (Routledge, 2020), *The Nature and Method of Economic Sciences*, includes a chapter on the nature of the economy, a topic that is also present in a chapter of an earlier book, *Philosophy of the Economy* (Springer, 2013). These chapters intended to deal with the economy in a broad sense, but, in fact, they discussed the nature and characteristics of economic actions. The work on the identity of the economic agent carried out by my advisor for my Ph.D. in economics, John B. Davis, has driven me to delve into this topic—the agent of economic actions—from the point of view of her actions and from the standpoint of the economic agent as a worker (a topic that I had partially developed as a chapter of the 2013 book). However, these advancements remain at a micro-level. Consequently, I also started asking myself about the nature of the macro-level and its relationship with the former. Another great component of the economy that deserved an inquiry into its nature was money. This has been my research subject for the past year (2020).

Then something happened. It dawned on me that I had explored some "components" of the economy: economic actions, the economic agent, workers, the nature of macroeconomic entities, and money. Thus, I decided to review my works on those topics, bringing them together with an ontological analysis. This is the path followed in the present book—it completes the review of the nature of economic sciences with the light shed by my earlier ontological studies. I proposed the book to Elizabeth Graber, Commissioning Editor of Economics at Palgrave-Macmillan, who enthusiastically supported the idea, and, after some interactions the proposal was approved. Elizabeth was promoted to Senior Editor on Sociology and Anthropology, and I continued working with Ruth Jenner, Commissioning Editor of Economics based in London, and Meera Seth. Geetha Chockalingam, along with Meera and Ruth, efficiently took the responsibility of the project coordination. I am very grateful to all of them.

My thanks also go to the people who read and commented on the chapters of this book. John Davis, as always, contributed with insightful remarks on Chapters 2, 4, and 8, respectively, dealing with metaphysical topics, the nature of the economy, and micro-macro relations. I presented a first version of the latter chapter at a seminar organized by the Centre for Humanities Engaging Science and Society (CHESS) at Durham University (October 14, 2020). I got valuable feedback from Nancy Cartwright, Peter Vickers, Richard Williams, and Omar El-Mawas. Another version was presented at the 15th Biennial Conference of the International Network for Economic Method, November 12–14, 2021, Arizona State University, Tempe, Arizona + Virtual, and at the LVI Annual Meeting of the Asociación Argentina de Economía Política (*The Argentine Political Economics Association*), November 17–19, 2021. I also met Daniel Heymann and Juan Carlos de Pablo, who provided valuable ideas about this chapter. Chapter 4 was presented at the XXVII Meetings of Economic Epistemology (Universidad de Buenos Aires, September 2021) and at the History of Economic Society Annual Meeting, Utrecht, December 2021. Robert Gallagher thoroughly commented Chapters 2 and 3. Geoffrey Hodgson read almost the whole book and proposed a modification of the title that I adopted.

J. F. Franck commented Chapter 5 on the identity of the economic agent. Chapter 6 touches on my work with my Ph.D. candidate (now Doctor) Omar Rodríguez and my co-author Belén Mesurado. Concerning Chapter 7 on the ontology of social collectives, I received useful remarks from Ivana Anton Mlinar, Frank Hindriks, Giancarlo Ianulardo, Dave Elder-Vass, Fin Collin, Brian Epstein, and Deborah Tollefsen. I presented an early version of this chapter at the 13th Biennial Collective Intentionality Conference, University of California, San Diego (online, August 16, 2021), and Matthew Baddorf, Giulia Lasagni, and Baris Kastas offered some comments and asked questions. I got interesting remarks about Chapter 9 on money from Alejandro Vigo, John Davis, Julián Giglio, Wade Hands, Rodrigo Laera, Carlo Natali, Fernando Tohmé, Eduardo Scarano, and three anonymous referees. I also presented a previous version at a Conference in Universidad de Buenos Aires and at the LV Annual Meeting of the Asociación Argentina de Economía Política. I am very grateful for all these contributions. As always, the entire responsibility lies solely on me.

I must acknowledge the use of some material and conceptual elements drawn from previously published works—specifically, from Chapters 2 and 9 of *Philosophy of the Economy. An Aristotelian Approach*, Springer, Dordrecht, 2013; "Two Conceptions of Economics," *Journal of Applied*

Economics, XIV/2, 2011, pp. 181–197; Chapter 2 of *Theoretical and Practical Reason in Economics. Capacities and Capabilities*, Springer, 2013; "Neurosciences, Neuroeconomics, and Metaphysics," in P. A. Gargiulo and H. L. Mesones (eds.), *Neurosciences and Psychiatry Update: Bridging the Differences*, Springer, Dordrecht, 2015, pp. 39–48; "Aristotelian Hylomorphism: A Framework for Non-Physicalist Philosophers About Philosophy of Mind," in Pascual A. Gargiulo and Humberto L. Mesones (eds.), *Psychiatry and Neuroscience Update. Vol. II: A Translational Approach*. Dordrecht: Springer, pp. 37–46; "Identity Theories in Economics: A Phenomenological Approach," co-authored with Ivana Anton Mlinar, in Peter Róna, László Zsolnai and Agnieszka Wincewicz-Price (eds.), *Words, Objects and Events in Economics: The Making of Economic Theory*, Springer, 2021, pp. 193–211; "Calling. Making the World a Better Place from Within Multinational Corporations," *Current Psychology*, co-authored with Belén Mesurado and Omar Rodríguez, 38/3, 2019, pp. 821–828, and "On Money as a Conventional Sign: Revisiting Aristotle's Conception of Money," *Journal of Institutional Economics*, 17/3, 2021, pp. 459–471. I appreciate the permissions granted by Elsevier, Springer Nature, Oxford University Press, and Cambridge University Press to use parts of these materials when it proved necessary, and I am also grateful to the anonymous reviewers of these published sections, who offered valuable advice.

I also want to convey my sincere gratitude to the academic institutions where I work for their support in my endeavors: IAE Business School (Universidad Austral, Pilar, Argentina); School of Philosophy, Universidad Nacional de Cuyo (Mendoza, Argentina), and Argentina's National Scientific and Technical Research Council (Consejo Nacional de Investigaciones Científicas y Técnicas, CONICET, by its initials in Spanish).

Lastly, it is only fair that my deepest gratitude goes to my family, to whom I dedicate this book. It would not have been possible without their support.

Buenos Aires and Mar del Plata, Argentina
January 2022

Ricardo F. Crespo

Contents

About the Author

Ricardo F. Crespo is Professor of Philosophy of Economics at IAE, Universidad Austral, and Researcher at the Argentine Council of Scientific Research. He is a graduate in economics and philosophy and earned a Ph. D. in Philosophy (Universidad Nacional de Cuyo, Argentina) and another in Economics (University of Amsterdam). He has extensively published articles and chapter books on his research topics. Recent publications of him include articles in *Synthese*, the *Journal of Institutional Economics*, *Foundations of Science*, *Cambridge Journal of Economics*, and the *Journal of Applied Economics*. His last book is *The Nature and Method of Economic Sciences: Evidence, Causality, and Ends* (Routledge, London, 2020). His current research interests include explanation in the social sciences, economic rationality, and ethics in economics.

CHAPTER 1

Introduction: Does Economics Deal with the Economic Stuff? Or Is the Economic Stuff Explained by Economics?

> But from time to time it is probably necessary to detach oneself from the technicalities of the argument and to ask quite naively what it is all about. (Friedrich A. von Hayek, 'Economics and Knowledge,' Economica 4/23, 1937, p. 54)

> It is essential that we look occasionally at the map or model for scientific progress that each of us surely carries around, consciously or unconsciously, in his head. (James M. Buchanan 1987, p. 21)

> Whenever there is some significant, institutionalized, and intellectually intriguing or perplexing human activity in place in society, it is regularly accompanied by a higher-order activity that has the task of reflecting on the lower-order activity. (Uskali Mäki 2021, p. 3)

Physics is the science dealing with the physical world. Physical laws formulated by physics effectively apply to the physical world almost perfectly. The way physics works corresponds to the nature of physical reality. We can, for example, plan astonishing travels to Mars with amazing exactness.

This is not the case of the economy and economics: plans designed by economic theorists often fail. The paradigmatic proof of the mismatch

R. F. Crespo, *The Nature of the Economy*,
https://doi.org/10.1007/978-3-031-02453-5_1

between economic reality and standard economics is provided by experimental and behavioral economics, which prove that, in specific situations, most people behave in the opposite way that standard economics predicts. Unlike physics, it seems that the way economics works not always corresponds to the nature of economic affairs. The aim of this book is to investigate the nature of these matters to adjust economics to it.

Standard economics' "rationality" frequently does not apply. Peter Hammond defines standard economics rationality as preference maximizing by consumers and profit maximizing by firms (1997: 31). He adds, "There seems to be little evidence of rationality in actual behaviour. This is true even for special classes of experimental subjects like economics or business students, who are supposed to understand something of what it means to be rational" (1997: 32–33). In this book, I will use the term "standard" economics as Hammond does. The theories of rationality that standard economics usually adopts are the Rational Choice Theory (RCT) and the Expected Utility Theory (EUT), which rely on the notion of instrumental rationality. Just as Dani Rodrik (2015: 7) uses the term "economics," I use the term "standard" to mean a discipline focused on methods, a way of doing social science associated with formal modeling and statistical analysis.

Sciences start with principles, which are axioms or conclusions of other, more basic sciences. For example, developments in medicine are based on conclusions drawn from biology, chemistry, and physics. Rationality is a first principle of economics—but not just any form of rationality: only the forms mentioned above, which are forms of instrumental rationality. This means rationality construed as people's "logical" or reasonable behavior, or, as Catherine Herfeld (2021: S3333) describes it, "human agents behave reasonably, given the external circumstances they face." It is almost trivial, and, hence, it does not serve as grounds for explanations or predictions. I have proposed to refer to this principle as "metaphysical," as opposed to a specification of it that I have called "empirical," adopting a meaning of the term "metaphysics" that I will not use in this book: metaphysical principles, according to my former use of the term, are not falsifiable principles, while empirical ones are (Crespo 2013).

The rationality principle specified by "standard economics"—RCT and EUT—is frequently falsified by empirical evidence.[1] Economics has

[1] For a more detailed analysis on the specifications of the rationality principle, see C. Herfeld (2021: S3334ff.) and see Herfeld (2020) for the different versions of RCT.

increasingly become empirical. However, these first principles, though often falsified, have remained untouchable. As Kevin Hoover has recently suggestively stated,

> Contemporary economics is pace Robbins and Mill favorable to empirical research and to the feedback from empirical results to economic theory. However, the Robbinsian first principles themselves are not the target for any revision within mainstream economics, but are held immutable as something like a Lakatosian hard core. Recalcitrant evidence may result in a revision of the details of the structure of constraints hypothesized in a problem, but is not allowed to weaken the commitment to the framework of constrained optimization. (2021: S3322)

In the same vein, Herfeld (2021: S3337) states that "[d]espite recent developments in behavioural economics, either the general variant of the rationality principle or its special status is rarely questioned by economists," and she provides examples of this fact taken from contemporary introductory textbooks on economics.[2]

While physicists largely agree on physical theory, there are numerous economic theories, which is paradoxical given that they are built on the same principles. Many philosophers of economics attribute this difference between physics and economics to the greater complexity of the economic realm. Yet, when one studies the complexity of physical reality, one starts questioning this argument. It must be something more than a greater complexity that explains this difference. There should be some radical, ontological, differences between the nature of physical and economic reality that explain it. Or complexity has a different nature than physical complexity. As also Herfeld states (2021: S3344),

[2] George Chorakakis explains, "The neoclassical research program (NRP) in economics, namely the meta-theory that rests on the axiomatic assumptions of rational choice, individualistic utility or profit maximization and ex ante equilibration of aggregate demand and supply, has shown extraordinary resilience vis-à-vis epistemic anomalies identified in early days. These anomalies are either ignored or tackled with ad hoc extensions of its 'protective belt' of auxiliaries assumptions, leaving its 'hard core' intact, and effectively extending its life and safeguarding its dominant position in academia instead of enhancing its empirical content or its explanatory power" (2020: 240).

> The complexity of human motivation, the multiplicity of motives that lead people to act, and the complexity of the environment within which they act imply that the general version of the [rationality] principle is never realized in an undisturbed and pure way.

It is a pressing demand to ascertain these differences. The result of this exploration might be that we must rest content with approximate results in economic theory and that we will need to circumscribe our investigations to specific contexts to gain certitude for our conclusions. This would be an important step, because we would advance our awareness of the limits of economic theories, and this could drive us to be more prudent in our economic plans—when a lot of people are affected by them.

Ultimately, the root of the problem lies with the first principles of economics. Who should provide these principles? Let us hear from Aristotle (*Metaphysics* IV, 3, 1005b 5–11):

> Evidently then it belongs to the philosopher, i.e. to him who is studying the nature of all substance, to inquire also into the principles of syllogism. But he who knows best about each genus must be able to state the most certain principles of his subject, so that he whose subject is existing things qua existing must be able to state the most certain principles of all things. This is the philosopher.

Thus, at its final stage, the culpability for this mismatch lies not with economists but with philosophers. There is a problem in the specification of the rationality principle that has to do with philosophical ideas. Rather imperceptibly, philosophy influences science and life. Maynard Keynes, who was above all a philosopher, had a clear conviction about this, which he somewhat expressed in the famous last paragraph of his *General Theory* on the influence and power of intellectual ideas for good or evil. Standard economic theories are the offspring of the ideas conceived by past centuries' philosophers, like David Hume or Immanuel Kant, and of the prevailing metaphysic physicalist worldview. This statement may seem to oversimplify the problem. Yes, these phenomena are complex, but the three reasons mentioned—Hume's and Kant's thinking and the physicalist worldview—explain a significant share of the mismatch.[3]

[3] The central argument of Nitasha Kaul's (2008) book is that economics is a kind of modernist knowledge and a by-product of Modern Enlightened Philosophy.

First, standard economic rationality, instrumental maximizing rationality, is an offspring of Hume's idea of rationality. In fact, Hume's writings have been called by John Hill Burton, one of his nineteenth-century biographers, "the cradle of political economy" (Burton 1846: 354). Hume downgraded the relevance of practical reason, reducing it to a technical function: designing the best way to achieve the goals defined by passions (see Hume [1739–1740] 1968: 415-II, iii, 3). Defining ends is no longer a role of reason. Consequently, there are no true or wrong ends or preferences. This is economics. Indifference curves are given. Economists' role is to adapt the budget to those curves in the best possible way, assuming that they "behave properly" (they have a negative slope; they are convex; they do not intersect, etc.)—a technical task. However, the economic agent does not behave that way: actual indifference curves do not behave properly. They might or not… it all depends. All economists can say about "abnormal" indifference curves is, as a friend economist told me, that "they are not useful to build a theory of demand for the real world." Yet, one may harbor some doubts about the reality of this world…

Christine Korsgaard remarks that "the limitation of practical reason to instrumental reason does not only prevent reason from determining ends; it even prevents reason from ranking them" (2001: 104). This is why Hume famously affirms that "It is not contrary to reason to prefer the destruction of the whole world to the scratching of my finger" ([1739–1740] 1968: 416). Accordingly, economics cannot correct such preferences. This is the example offered by Terence Irwin: "If I could book a flight from New York to London on one airline for $600 and on another airline for $700 and airlines are otherwise equal, reason, as Hume understands it, is indifferent between them" (2007–2009: Volume II, 585). That is, if the traveler chooses to buy the $700 ticket, economics can only assert that, from an economic rationality perspective, this choice is irrational. However, there may be other reasons for the traveler to choose "irrationally," and buying a ticket for $700 is an economic action. Herfeld (2021) proposes we "consider the rationality principle as a functional a priori principle along the lines of a pragmatic theory of constitutive principles in science" (2021: S3356). Another possibility is to broaden the notion of rationality, reconsidering practical rationality not reduced to instrumental rationality as a principle for economics combined with the latter.

Second, standard economics' dismissal of the realism of assumptions, paradigmatically expressed by Milton Friedman (1953), is an offspring of Kant's transcendental philosophy. Kant declared that Hume had woken him from his "dogmatic slumber." He proposed infusing a Copernican revolution into reason. As described by Anthony Kenny, "Instead of asking how our knowledge can conform to its objects, we must start from the supposition that objects must conform to our knowledge" (2006: 102). As I will explain in the next chapter, this Copernican revolution perspective backfires, because it conditions the possibilities of knowing the "something" that lies behind sensorial appearances "of which we know nothing at all nor can know anything in general (in accordance with the current constitution of our understanding)" (*Critique of Pure Reason* A 250, 1999: 348). Hence, we do not know the ultimate reality—we just provide a structure for the appearances in our mind. "Reason has insight only into that which it produces after a plan of its own" (*Critique of Pure Reason* B xiii, 1999: 109). The consequence for economics is that we can only aspire to offer an explanation through models that predict a convincing explanation of facts—i.e., "to save the phenomena"—but we do not know if we are accessing the very nature of these phenomena.

Finally, the third reason stems from the predominant metaphysical view or *Weltanschauung*. Metaphysics—construed as a worldview—is always present in science. This idea has been developed and consolidated from Pierre Duhem to the present, through Karl Popper, Hanson, Quine, Kuhn, among others.[4] Philosopher of science Evandro Agazzi (1988: 19) states:

> Science [...] cannot be pursued without one's using certain criteria of intelligibility which are prior to the specific tasks it involves. In fact, every advancement of some science which has been presented as a "liberation from metaphysics" has actually been tantamount to discarding a particular metaphysical framework and accepting (often unconsciously) a different one [...] Therefore it is much more reasonable to be aware of the metaphysics one has, rather than have a metaphysics without knowing it.

[4] I explored this evolution in an earlier work (2015), from which I draw some material for this chapter. However, this dependence of science on the physicalist worldview should not be exaggerated (see Dilworth's criticism 1994: 138ff.).

His final advice is useful to spot the mismatch between economics and the economy. Today's the metaphysical view that impregnates science is physicalism. Craig Dilworth (2006) underscored that modern science applies specific, fundamental metaphysical principles which produce a physicalist, deterministic (though not rigid) view of reality that is not accepted without reservations by all scientific disciplines, depending on their subjects. Physicalism is the metaphysical worldview that everything in the world is ultimately physical. It correlates with the so-called thesis of the "causal closure of the physical world," which states that all causes and effects in the physical realm are physical. David Papineau explains the consequences of this idea:

> the causal closure of the physical does give rise to a powerful argument for reducing many prima facie non-physical realms to physics: for it indicates that anything that has a causal impact on the physical realm must itself be physical. This is because the causal closure of the physical seems to leave no room for anything nonphysical to make a causal difference to the physical realm, since it specifies that every physical effect already has a physical cause. (2009: 52)

This thesis bears a strong impact on the discussions about the relationship between mind and body. Nonetheless, it also bears an impact on social sciences, including economics. The economy is a human reality in which things like the mind, will, desires, and freedom play a leading role. Dilworth notes that physicalism introduces a tension in social sciences and particularly in economics:

> [T]here is a particular tension in the economist's conception of human nature. On the one hand, the notion of free will is integral to it, since without free will the rationality principle would make no sense. On the other hand, however, no economic actor has the freedom not to follow the rationality principle, which itself determines how he or she is to act. (2006: 135)

This tension does not only apply to standard economics. My book *Economics and Other Disciplines. Assessing New Economic Currents* (2017) looks at the new "reverse imperialist" economic theories to ascertain whether they have been influenced by this physicalist worldview. The conclusion corroborates that they have indeed been influenced by it, but that a tension remains between this physicalist worldview and the defense of freedom, creativity, novelty, values, and morality.

Let me explain what "reverse imperialist" economic currents are and why they are so-called. We need to know their opposites in order to understand them. Starting approximately in the 1980s or even before, standard economics has been "exporting" its logic—instrumental rationality—to other social sciences. In his article "The Expanding Domain of Economics," Jack Hirshleifer approvingly describes this process that has been known as "economic imperialism" (see also Lazear 2000):

> *There is only one social science.* What gives economics its imperialist invasive power is that our analytical categories—scarcity, cost, preferences, opportunities, etc.—are truly universal in applicability. Thus economics really does constitute the universal grammar of social science. (1985: 53, italics in the original)

This invasion is not at all "pacific." As George Stigler asserts, "economics is an imperial science: it has been aggressive in addressing central problems in a considerable number of neighboring social disciplines and without any invitations" (1984: 311). Ronald Coase (1978: 207) explains and complains about the double tendency associated with this process:

> The first consists of an enlargement of the scope of economists' interests so far as subject matter is concerned. The second is a narrowing of a professional interest to a more formal, technical, mathematical analysis. This more formal analysis tends to have a greater generality. It may say less, or leave much unsaid, about the economic system, but, because of its generality, [...] economics becomes the study of all purposive human behavior, and its scope is, therefore, coterminous with all of the social sciences.

Gary Becker's (1976, 1993) economic approach best exemplifies this perspective. The reductionist tendency of standard economics is actually an obstacle to the introduction of new ideas and approaches, thus constituting a kind of "protectionism" from empirical data and the possible contributions of other social sciences, as Stefano Zamagni notes (2021: 14 and passim).

The above-mentioned mismatch between the economy and economics as well as the distortion of the nature of the subject matters of other sciences produced by the conclusions of the "imperialist" tendency is slowly making room for a reaction embodied in this new process called "reverse imperialism." This new process consists in moving in the opposition direction, importing the *rationale* and other inputs of other social sciences into economics. Thus, a "mainstream pluralism" is emerging (Davis 2008, 2011).

However, the question remains, are these new currents escaping from the physicalist *ethos*? As stated above, they escape only partially. My book (2017) concludes that the impact of physicalism is strong in neuroeconomics, behavioral economics, some evolutionary theories, and hedonistic happiness economics. Concerning institutional economics, its impact depends on the concept of agency, habits, and institution adopted. Some new happiness theories adopting a richer concept of happiness than the hedonic, as well as the capability approach and civil economy theories stay outside the physicalist umbrella. This conclusion is somewhat paradoxical: physicalism partially remains, but is it an adequate perspective to analyze the subject matter of economics or not? It all depends on the nature of the economy.[5]

This pluralism of economic theories can prove positive if we have a clear understanding of the nature of its subject matter—the economic stuff, which possesses several dimensions. Are economic decisions and actions mere technical means to achieve some preferences? Discovering the nature of a thing is a metaphysical task that must be undertaken, because the meaning and nature of the economy and its fundamental components is a still unclear topic. Economists do not even share a consensus about the meaning of the economy. We need to know the nature of the economy to deal with it appropriately. As Israel Kirzner puts it, "Definitions of economic science have time and again required preliminary discussions revolving around the question whether the discipline concerned a kind of object, a kind of activity, a kind of man, or a kind of satisfaction or welfare" (1976: 6).

As Mäki (2021) explains these kinds of questions have to be answered by philosophy of economics because "these are questions that the lower-order activity itself [economics] does not customarily or regularly ask and answer in its ordinary mode" (2021: 3). A common purpose of the

[5] Geoffrey Hodgson, in a very philosophical book, complains about the inadequacy and misleading character of the physicalist metaphors underlying economic analysis (2015: 3, 18, 166–167, 266ff.).

higher-order reflection, Mäki states, "is to put forth a *programmatic statement*, an organized set of principles and guidelines telling how to do economics properly or better" (2021: 5).

Eric Schliesser (2015) has drawn our attention to an old criticism by Joseph Cropsey (1955) of the New Welfare Economics (NWE). Schliesser remarks that Cropsey was aptly qualified for this task since he had recently earned his Ph.D. in Economics from Columbia University, where the NWE was blossoming (2015: 848). Schliesser (2015: 847) also recalls that George Stigler (1943) had previously argued that this branch of economics adopted a question-begging consensus over society's values. For Cropsey, the NWE assumes metaphysically wrong conceptions underlying its mathematical analysis, its psychological assumptions, and its implicit political or institutional framework. Consequently, Cropsey affirms, "the entire structure of modern welfare economics rests upon a sandy foundation" (1955: 124). Thus, he states, we need to undertake its reconstruction upon a plan based on a substantive conception of human nature: "every logic presupposes a metaphysics" (1955: 118). Indeed, I agree with Hasse Ekstedt when he states that "In economics one has the feeling that the urge to understand and explain the short-run economic reality has overshadowed the important analysis of the used scientific concepts" (2013: 1).

Philosophy is useful for economists (and all scientists) because it does the prior work of discerning the nature of their subject matters. As Aristotle states in *Metaphysics*, "For those who wish to get clear of difficulties, it is advantageous to discuss the difficulties well; for the subsequent free play of thought implies the solution of the previous difficulties" (III, 1, 995a26–29). Philosophical definitions help to prevent mistakes as well as to save time by pointing directly to the kind of research that is worth doing and to adequate methodologies.[6] I am one of those philosophers of economics that thinks "that practicing economists would benefit from paying attention to our work" (Vromen 2021: 27).

Thus, the aim of this book is to ascertain the metaphysical nature of the "components" or "constituents" of the economy in order to appraise the epistemological statute that economic science should adopt. The second chapter defines the metaphysical theory that will be used throughout the book. There are different metaphysical theories, and I explain why I have

[6] However, as Hodgson states, "Lamentably few social scientists these days have a solid grounding in philosophy, including the philosophy of their own discipline" (2015: 10).

adopted the Aristotelian version of metaphysics. Chapter 3 introduces the metaphysical categories that will be applied to the components analyzed in this book. These categories embody the way of being or nature of entities. Aristotle establishes specific categories, and I have added the categories used by three scholars who also undertake a metaphysical analysis of the economy: Nancy Cartwright, Tony Lawson, and Uskali Mäki. Chapter 4 deals with the economy in general, actually focusing on individuals' economic decisions and actions. Aristotle, Cartwright, Lawson, and Mäki take the stage once again. Chapter 5 tackles the topic of the nature of the economic agent—a topic I have dealt with in two previous works: one written alone and the other co-authored with Ivana Anton Mlinar. Chapter 6 considers the economic agent from a different perspective, as a worker. There are precedents in two works—one I wrote by myself and another one co-authored with Belén Mesurado and Omar Rodríguez. Chapter 7 on social ontology serves as a preamble to Chapter 8 on the nature of macroeconomic entities and their relationship with microeconomic entities. Chapter 9 deals with the nature of money, and it is based on my 2021 article on the topic. Yet, I have introduced some significant changes. I thank Alejandro Vigo for his comments that prompted them. Finally, once the previous metaphysical characterizations of the components of the economy have been completed, I turn back to the nature of economic science itself.

I should make it clear from the start that not all the constituents of the economy will be analyzed: hence the subtitle of the book, "Aristotelian Essays on the Philosophy and Epistemology of Economics," which hints at its incomplete character. Some relevant pieces will be missing, including but not limited to the nature of the market, economic value, or capital. This work shall be undertaken in future investigations.

References

Agazzi, E. (1988). "Science and Metaphysics: Two Kinds of Knowledge", *Epistemologia. An Italian Journal for the Philosophy of Science* 11/11: 11–28.

Becker, G. (1976). *The Economic Approach to Human Behavior*. Chicago: Chicago University Press.

Becker, G. (1993). "Nobel Lecture: The Economic Way of Looking at Behavior", *Journal of Political Economy* 101/3: 385–409.

Buchanan, J. M. (1987). *Economics. Between Predictive Science and Moral Philosophy*. Texas: Texas A&M University Press.

Burton, J. H. (1846). *The Life and Correspondence of David Hume*. Edinburgh: William Tait.
Chorafakis, G. (2020), "Emergence Versus Neoclassical Reductions in Economics", *Journal of Economic Methodology* 27/3: 240-262.
Coase, R. (1978). "Economics and Contiguous Disciplines", *The Journal of Legal Studies* 7/2: 201–211.
Crespo, R. F. (2013). "Two Conceptions of Economics and Maximisation", *Cambridge Journal of Economics* 37: 759–774.
Crespo, R. F. (2015). "Neurosciences, Neuroeconomics, and Metaphysics". In: P. A. Gargiulo and H. L. Mesones (ed.) *Neurosciences and Psychiatry Update: Bridging the Differences*. Dordrecht: Springer, pp. 39–48.
Crespo, R. F. (2017). *Economics and Other Disciplines. Assessing New Economic Currents*. London: Routledge.
Crespo, R. F. (2021). "On Money as a Conventional Sign: Revisiting Aristotle's Conception of Money", *Journal of Institutional Economics*.
Cropsey, J. (1955). "What Is Welfare Economics?" *Ethics*, 65/2: 116–125.
Davis, J. B. (2008). "The Turn in Recent Economics and the Return of Orthodoxy", *Cambridge Journal of Economics* 32: 349–366.
Davis, J. B. (2011). *Individuals and Identity in Economics*. Cambridge and New York: Cambridge University Press.
Dilworth, C. (1994). *Scientific Progress*, 4th Edition. Dordrecht: Springer.
Dilworth, C. (2006). *The Metaphysics of Science*, 2nd Edition. Dordrecht: Springer.
Ekstedt, H. (2013). "The Mysterious Money—Some Epistemological Notes", 10th International Conference on Developments in Economic Theory and Policy, Bilbao, June 27–28, 2013.
Friedman, M. (1953). "The Methodology of Positive Economics". In M. Friedman (ed.) *Essays in Positive Economics*. Chicago and London: University of Chicago Press, pp. 3–43.
Hammond, P. J. (1997). "Rationality in Economics", *Rivista internazionale di Scienze sociali* CV: 247–288, https://web.stanford.edu/~hammond/ratEcon.pdf. Accessed 20 July 2021.
Herfeld, C. (2020). "The Diversity of Rational Choice Theory: A Review Note", *Topoi* 39: 329-347.
Herfeld, C. (2021). "Understanding the Rationality Principle in Economics as a Functional a Priori Principle", *Synthese* 198 (Suppl 14): S3329–S3358.
Hirshleifer, J. (1985). "The Expanding Domain of Economics", *American Economic Review* 75/6: 53–68.
Hodgson, G. M. (2015). *Conceptualizing Capitalism. Institutions, Evolution, Future*. Chicago and London: University of Chicago Press.
Hoover, K. (2021). "First Principles, Fallibilism, and Economics", *Synthese* 198 (Suppl 14): S3309–S3327.

Hume, D. ([1739–1740] 1968). *A Treatise of Human Nature*, ed. L. A. Selby-Bigge. Oxford University Press (reprinted).

Irwin, T. W. (2007–2009). *The Development of Ethics: A Historical and Critical Study*. Oxford: Oxford University Press.

Kant, I. ([1787] 1999). *Critique of Pure Reason*, translated and edited by Paul Guyer and Allen Wood. Cambridge: Cambridge University Press.

Kaul, N. (2008). *Imagining Economics Otherwise. Encounters with Identity/Difference*. Abingdon: Routledge.

Kenny, A. (2006). *The Rise of Modern Philosophy*. Oxford: Clarendon Press.

Kirzner, I. M. (1976). *The Economic Point of View. An Essay in the History of Economic Thought*, 2nd edition. Kansas City: Sheed and Ward.

Korsgaard, C. (2001). "Skepticism About Practical Reason". In: E. Millgram (ed.) *Varieties of Practical Reason*. Cambridge, MA and London: MIT Press, pp. 103–126.

Lazear, E. P. (2000). "Economic Imperialism", *The Quarterly Journal of Economics* 115/1: 99–146.

Mäki, U. (2021). "The Field: Tasks, Pasts, Futures", *Journal of Economic Methodology* 28/1: 3–13.

Papineau, D. (2009). "The Causal Closure of the Physical and Naturalism". In: B. McLaughlin, A. Beckermann and S. Walter (eds.), *The Oxford Handbook of Philosophy of Mind*. Oxford University Press, pp. 53-65.

Rodrik, D. (2015). *Economic Rules*. Oxford: Oxford University Press.

Schliesser, E. (2015). "On Joseph Cropsey's 'What Is Welfare Economics'," *Ethics* 125: 847–850.

Stigler, G. (1943). "The New Welfare Economics," *American Economic Review* 33: 355–359.

Stigler, G. (1984). "Economics—The Imperial Science?" *Scandinavian Journal of Economics* 86/3: 301–313.

Vromen, J. (2021). "What Are We Up to?" *Journal of Economic Methodology* 28/1: 23–31.

Zamagni, S. (2021). "Per una scienza economica non riduzionista", *Cum-Scientia* III/5: 11–34.

CHAPTER 2

Metaphysics, Ontology, and Metaphysical Theories

As explained in the Introduction, this book aims at discerning the deep meaning—the nature or essence—of the economy and of the fundamental elements in the economic realm. Its first goal is to determine what the economic entities are and to discover what their nature is. This is a philosophical task—more specifically, a metaphysical or ontological job. The second goal is to draw conclusions from the previous analysis on the appropriate way of learning about these entities and doing science with them—a task that falls within the purview of the branches of philosophy called epistemology and philosophy of science. In turn, the purpose of this chapter is to establish the meaning of metaphysics and ontology, as well as to introduce the metaphysical theory that will be adopted in the book.

The words "metaphysics" and "ontology" go back a long way. Andronicus of Rhodes (70 BC) used the phrase "*Ta meta ta physika*"—"the ones after the physical ones"—to entitle fourteen books written by Aristotle about what he called "first philosophy" or "theology." For Aristotle, this science deals with "being as such" and with the first causes and categories of beings or entities, such as substances or things, and accidents or properties (*Metaphysics*, IV 1003a 18–19). Rationalist philosopher Christian Wolff (1679–1754) coined the term "ontology" to broaden the subject matter of metaphysics in order to include matters like the relationship between mind and body, free will, and personal identity. However,

R. F. Crespo, *The Nature of the Economy*,
https://doi.org/10.1007/978-3-031-02453-5_2

the meaning of these terms has not remained unchanged, and there are numerous conceptions about them corresponding to different theories about the nature of reality and its knowledge.

For example, Columbia University Professor Achille Varzi (2011: 407) states:

> According to a certain, familiar way of dividing up the business of philosophy, made popular by Quine, ontology is concerned with the question of what entities exist (a task that is often identified with that of drafting a "complete inventory" of the universe) whereas metaphysics seeks to explain, of those entities, what they are (i.e., to specify the "ultimate nature" of the items included in the inventory).

Indeed, Quine describes the "ontological problem" with three words: "what is there?" (1948: 21). He considers "blurring the supposed boundary between speculative metaphysics and natural science" (1951: 20) as one of his philosophical undertakings. Michael Loux and Thomas Crisp explain, following the Aristotelian tradition, that "Metaphysics is the most general of all the disciplines; its aim is to identify the nature and structure of all that there is. Central to this project is the delineation of the categories of being. Categories are the most general or highest kinds under which anything that exists falls" (2017: xi)—that is, the "ultimate nature" in Varzi's words. As Agazzi puts it,

> two basic meanings of it [metaphysics] may be found already at the beginning of its official history (i.e., in the *Metaphysics* of Aristotle): on the one hand metaphysics is conceived as the science of 'reality as such,' i.e. of the most universal *features* of reality; on the other hand, it is conceived as the science of those *dimensions* of reality which overstep its empirically ascertainable level (or, to put it briefly, as the science of the 'suprasensible'). (1988: 11)

However, though using the same name, the meaning and reach of metaphysics according to Quine, on the one hand, as well as Loux and Crisp, and Agazzi, on the other hand, are completely different. As Tuomas Tahko affirms, "According to the Quinean approach, the key questions of metaphysics concern the existence of different kinds of things, whereas the Aristotelian approach focuses on the natures or essences of these things" (2011: 26).

Varzi clarifies that his Quine-based divide between the tasks of ontology and metaphysics "is everything but universal among philosophers. There are many other, different ways of understanding the terms 'ontology' and 'metaphysics', some of which can certainly claim a respectable pedigree" (2011: 407). He adds, "But never mind; I am not interested in defending the view or in criticizing it, as very little depends on it. I am citing it just to fix a certain distinction and to settle on a terminology" (2011: 407). That is also my intention. I will take Varzi's task division, with ontology defining the inventory of entities and metaphysics explaining the ultimate nature of entities. Yet, I draw away from Quine concerning his conception of ontology and metaphysics.[1]

Before proceeding to the inventory of economic entities, this chapter intends to define the metaphysical theory that will be adopted, while the next chapter will establish the categories of being that will be used in the book to analyze the nature of economic "entities" or "beings" (I use the words "entity" and "being" without any specific metaphysical commitment just yet).

The next steps in subsequent chapters will be the following: first, Chapter 4 will tackle two questions—"what is the nature of the economy?" in metaphysical terms, and "which are the economic entities?" from an ontological standpoint. Next, the successive chapters will delve into the metaphysical investigation of the nature of those economic entities.

Two Metaphysical Theories

There are different theories about categories of being. Let me first relay my personal experience, which proves relevant as it led me to the theory that I will adopt here. As I noted in the acknowledgments, my father was an economist, and, during my youth, we talked a lot about economics and the nature of economic reality. These conversations impressed an

[1] "Quine [Willard van Orman Quine] himself does not much use the word "metaphysics." He puts forward views on metaphysical topics—such as ontology, time, and abstract objects—but within the confines of his naturalism" (Hylton and Kempt 2020). Still, some scholars wonder whether Quine was a "metaphysician," equating this word with "ontologist." I prefer to maintain the difference because it is one thing to detect what there is (considering this as an ontological task) and another, to ascertain its nature (a metaphysic task), which Quine does within a disruptive context as regards classical metaphysics.

early vocation for the philosophy of economy upon me. While they sometimes use very abstract models, economists are largely realists: they work with decidedly down-to-earth realities. They need concepts that correlate with facts—otherwise, their theories fail. This is probably why, when I studied philosophy, the classical Aristotelian metaphysical approach appealed to me: his proposed categories of beings convincingly describe the elements in the existing world. This is also probably why I was surprised when my professor of the History of Modern Philosophy explained Kant's "metaphysical" theory, i.e., his "transcendental philosophy." From an Aristotelian point of view, the expression "Kant's metaphysics" is an oxymoron because it actually entails the rejection of classical metaphysics—or it refers to another version of metaphysics.

Kant thinks that Aristotle's metaphysics exceeds the possibilities of human reason. For Kant, theoretical reason cannot penetrate the nature of things. Kant refers to the nature or essence of a thing as its *noumenon*—that is, "a thing which must be cogitated not as an object of sense, but as a thing in itself [*Ding an sich*]" (*Critique of Pure Reason*, B310, 1999: 362). As he puts it, "we have no insight into the possibility of such *noumena,* and the domain outside of the sphere of appearances is empty (for us)" (*Critique of Pure Reason*, A255, 1999: 362). We cannot reach the *noumena* with our theoretical reason. Agazzi elaborates:

> As the result of his "critical" investigations, he [Kant] was convinced that metaphysics was admissible as a "science" in its (suitably reformulated) sense of being the doctrine of the most universal features of reality, while it could not be credited with the qualification of "science" in its second sense, i.e. in the sense of being the knowledge of a suprasensible dimension of reality. The suitable reformulation just mentioned was that metaphysics may be seen as the doctrine of the most universal *features* of our knowledge, i.e. of those features which make knowledge at all possible *a priori*, and which determine in such a way the whole domain of the *objects* of knowledge (the world of *phenomena*). Instead of thinking of metaphysics as the science of the universal and necessary principles of "what exists," he confined it to being the science of the universal and necessary principles of "what is knowable." The second traditional part of metaphysics—that of being the science of the suprasensible—was displaced from the status of "knowledge" to the status of "faith". (1988: 11)

For Kant, the "pure intuitions" of time and space bring a structure to appearances, "the raw material of our sensuous impressions" (*Critique of*

Pure Reason, Introduction B1, 1999: 136), "a chaos of sense impressions"—an expression used by Friedrich Nietzsche to refer to Kant's theory in an even more extremist anti-metaphysical way (Nietzsche [1901] 1968, §569: 307). Then, once this process is achieved, Kant maintains that the imagination mediates between sensation and understanding. The latter, he upholds, applies twelve "categories" a priori—that he deduced (a "metaphysical" deduction) from judgment forms—to the resulting previous *schema*, thus providing a structure to it. However, this structure is no more than our organization of the objects of senses: it does not tell us anything about the "thing in itself." As Gilson and Langan elaborate, "The 'being' of the things of our experience is seen to reside, not in the principles within them, but in the power of representation on the part of the subject making them *to be for us*" (1963: 423). Also, as Paul Guyer and Allen Wood explain in the Introduction to the Cambridge Edition of the *Critique of Pure Reason*, "We cognize these things not as they are themselves but only as they appear under the conditions of our sensibility" (1999: 8). This is why Kant views the elements of cognition—and not reality—as the subject matter of metaphysics: "things as they are in themselves ('noumena') might be *thought* but not *known*" (Guyer and Wood 1999: 10). Henry Veatch explains Kant's and Kantian-oriented philosophers' position about the reach of their notion of knowledge:

> a transcendentally justified knowledge can only be a knowledge of appearances and not of reality or of the thing in itself. And the reason is that, in the very nature of the case, what thus comes to be justified in a transcendental justification are but our own human schemes or conceptual frameworks, or the ways in which, through our various theories or hypotheses, we human beings have come to order things and to put them together and thus, in a quite literal sense, to have placed a certain construction upon them. Consequently, the knowledge of things that we attain by such means can only be a knowledge of things as they appear to us to be, in the light of the theories and hypotheses through which we view them and come to regard them, and hence not necessarily a knowledge of things as they really are in themselves at all. (1978: 416–417)

This view bears an impact on science. As Alfred Whitehead (1929: 60) points out, "Kant drove a wedge between science and speculative reason," because science is about phenomena, not about unknowable (for Kant) *noumena*. Then, reason does not perform a "discovery," but rather it plays a "constructivist" role, building the object of knowledge through a priori mental categories and judgments. Guyer and Wood explain, "The human understanding, therefore, is the true lawgiver of nature, and the successes of modern science are due to its conduct of its inquiries in accordance with a plan whose ground lies a priori in the structure of human thought" (1999: 21). Science ultimately relies on the very reason that builds it.[2]

I agree with Loux and Crisp's argument in favor of the Aristotelian theory over Kant's perspective (2017: 1):

> The case for this Kantian conception of metaphysics is not, however, particularly impressive; for if there are problems with characterizing the world as it is, there ought to be similar problems with characterizing our thought about the world.

In other words, the Kantian option leaves us in an agnostic situation. Thus, I prefer to adopt Aristotle's. I "naturally" lean towards this choice. I am not interested in knowing the structure of our knowledge, but the structure of economic reality. Obviously, our knowledge of economic reality conceptualizes it in a particular way, but I am more interested in the subject conceptualized than in the conceptualization itself. I see myself as part of the "philosophers who view metaphysics in pre-Kantian terms, tak[ing] metaphysics to have as its task an account of the nature and structure of the world itself" (Loux and Crisp 2017: 7), not of a structure imposed by my mind.

A relevant problem underlying the different metaphysical positions is the role of the intellect in knowledge (the subjacent theory of knowledge or epistemology). While Kant stresses its active role, Aristotle regards intellect as having a passive or receptive role in knowledge, but it certainly also plays an active role (cf. *On the Soul* III, 5 on the "active mind"). For Kant, the intellect does almost all in knowledge, whereas, for Aristotle, the combination between receiving and building is more balanced. Like most more modern thinkers, Kant believes epistemology takes priority, while Aristotle places metaphysics at the topmost. For Aristotle, science and its method must be adapted to the subject matter, instead of adapting the subject matter to science. This is impossible for Kant because, given

[2] This influential way of thinking has also impacted economics, introducing a tension between abstract modeling and the quest for knowing real facts. See, for example, Bob Goudzwaard and Harry de Lange (1995: 49–50) on the influence of Kant's thinking on economics.

that he cannot know the ultimate nature of reality, he must build the subject matter based on sensations of time and space, imagination, and intellectual categories.

I asserted earlier that Quine's metaphysics differed from Loux and Crisp's Aristotelian views on metaphysics. Against the backdrop of the denial of metaphysics by logical positivists in the early 1900s, Quine may seem to defend it with the following statement:

> if there is no proper distinction between analytic and synthetic, then no basis at all remains for the contrast which Carnap urges between ontological statements and empirical statements of existence. Ontological questions then end up on a par with the questions of natural science. (1966: 134)

Nonetheless, Huw Price asserts about this text:

> This sounds like good news for ontology, but actually it isn't. Quine's criticism of Carnap is in no sense a vindication of metaphysics. Of course not, for if all issues are ultimately pragmatic, there can't be the more-than-pragmatic issue of the kind the metaphysician requires. (1997: 7)

Let us go back to Quine himself (1953: 78–79):

> it is meaningless, I suggest, to inquire into the absolute correctness of a conceptual scheme as a mirror of reality. Our standard for appraising basic changes of conceptual scheme must be, not a realistic standard of correspondence to reality, but a pragmatic standard.

And Veatch comments (1978: 421), "On Quine's pragmatic standard, the legitimacy of any linguistic or conceptual scheme is not going to lie in any fancied correspondence that it may have to reality, but solely in its efficacy in facilitating what Quine likes to call 'communication and prediction'." That is, Quine replaces Kant's intuitions and categories as the criteria to structure experimental data with a pragmatic aim of making sense of the latter, but reality remains unknown. Veatch quotes Richard Cartwright who believes that, for Quine, "to enquire into the ontological commitments of a theory is not to ask what there is, but only to ask what that theory says there is" (1954). Veatch's conclusion in a carefully argued article is clear: Quine is "a covert or crypto-Kantian!" (1978: 408)—even more radical than Kant himself. Everyday judgments about the nature of ordinary objects and scientific knowledge are all only hypothetical. Quine adopts the Popperian criticisms not only of induction but also of falsationism. For Quine, scientific hypotheses run the same fate of every type of knowledge: they are only pragmatic devices to explain and predict, but they are blind when it comes to the nature of reality.

This notion has been and is present in economic science: it echoes most economists' often pragmatic view of theories and models: the crucial test for their success and the ruling criterion to choose them is pragmatic—prediction. However, while prediction is surely relevant, it is not always certain. This perspective implies putting the cart before the horse. I do not want to be told by science what the nature of economic events is for pragmatic reasons; rather, metaphysics should perform that task. I do not want to choose the subject matter for methodological reasons but to adapt the methodology to the subject matter.

Even economic terms share this spirit, as Marshall's views have revealed:

> Its [economics] reasonings must be expressed in language that is intelligible to the general public; it must therefore endeavour to conform itself to the familiar terms of everyday life, and so far as possible must use them as they are commonly used. ([1920] 1962: 43)

Robbins, however, has asserted:

> It is often urged that scientific definitions of words used both in ordinary language and in scientific analysis should not depart from the usages of everyday speech. No doubt this is a counsel of perfection, but in principle the main contention may be accepted. Great confusion is certainly created when a word which is used in one sense in business practice is used in another sense in the analysis of such practice (...) But it is one thing to follow everyday usage when appropriating a term. It is another thing to contend that everyday speech is the final court of appeal when defining a science. For in this case the significant implication of the word *is* the subject-matter of the generalizations of the science. (1935: 5, nt.)

I agree with Marshall, not with Robbins. Though not free from presuppositions, everyday speech is sometimes closer to the nature of economic entities than scientific generalizations. The economic methodology—maximizing instrumental rationality—often distorts the nature of the things studied. This is why it often fails and is criticized.

The thesis of this book is that we need to base economic knowledge on more solid grounds. The economy exists independently of economics. It is not the task of economics to define the economy—a philosophical task—but to study and explain it once defined and even to predict its evolution. The method must be established only once the nature of things has been ascertained; science's methodology must be adapted to the nature of its subject matter—not the other way around. Metaphysics comes first. Robert W. Batterman (2020: 77) explains:

> Reality is so complex and many-faceted that we are bound to treat it in different frameworks of consideration that address the very different aspects of its nature. Much will depend on whether the domain at issue is physics, or chemistry, or biology, or economics, or social affairs. The extent to which descriptive precision, phenomenal predictability, and explanatory generality can be realized will differ from field to field depending on the ontology and phenomenology of the domain.

Even within science we have different levels or kinds of entities: natural, social, and artificial, micro, meso, and macro. This ontological plurality calls for a methodological plurality. We cannot allow epistemology to determine ontology. Indeed, science needs to simplify things, but the epistemological simplification should not condition and actually become the ontological conception. We must start by discovering the different economic entities and establishing their metaphysical nature. As Carl Menger states in his *Investigations into the Method of the Social Sciences* ([1883] 1985), we must start by "taking account of the object by a separate investigation and going on to the definition of economics after the solution of the pertinent preliminary question" ([1883] 1985: 198, Appendix II, note 131).

The Categories of Being, Representationalism, and the Economy

Aristotle provides different definitions or aims for metaphysics, and the one that is relevant here is "the science of being qua being." Indeed, Aristotle's "first philosophy" or "theology" (later called "metaphysics") is "a science which investigates being as being and the attributes which belong to this in virtue of its own nature" (*Metaphysics*, IV, 1003b 31). As Loux and Crisp elaborate, "Metaphysics considers things as beings or as existents and attempts to specify the properties or features they exhibit just insofar as they are beings or existents" (2017: 3). These properties or features are ultimately the different forms of beings or the "categories" of being. While Aristotle regards these categories as deduced from the contemplation of reality, Kant believes they are deduced from human judgments. Amy Thomasson (2018) states:

> As Studtmann (2007) puts it, Aristotle "assumes rather than defends a posture of realism with respect to the metaphysical structures of the world." Given this approach, a complete system of categories would offer a systematic inventory of what there is, considered at the most abstract level.

Thomasson (2018) adds:

> it is clear that for Kant the categories find their original source in principles of human understanding, not in intrinsic divisions in mind-independent reality, and are discoverable by paying attention to possible forms of human judgment, not by study of the world itself, nor by study of our contingent manners of speaking.

Kantian philosophers argue that what we know are not things, but a result of the application of mind categories to phenomena, supposedly "representations" of things, not the things themselves. In fact, the "representationalist" theory of knowledge prevails among contemporary philosophers:

> Philosophers who take the notion of a conceptual scheme seriously will take metaphysics to be concerned with our way or ways of representing the world. [...] They agree with Kant that our thought about the world is always mediated by the conceptual structures in terms of which we represent that world [...] what I grasp is not the object as it really is independently of my thought about it. (Loux and Crisp 2017: 7)

However, an anti-representationalist tradition prevails among economist-philosophers like John Maynard Keynes, who advocates the ability of knowing the thing itself. In *A Treatise on Probability,* he talks about acts of penetration "into the real world" ([1921] 1973: 56), which he describes as direct knowledge, "intuition or direct judgment" ([1921] 1973: 56, 70, 121), "a faculty of direct recognition" ([1921] 1973: 57), "the hidden element of direct judgment or intuition" ([1921] 1973: 69), "intuitive power" ([1921] 1973: 76), and "direct judgment (…) by seeing" ([1921] 1973: 121). He speaks of "direct inspection" and "direct unanalysable intuition" in "My Early Beliefs" (1972: 437).

Is this direct knowledge a mere knowledge of representations? No, Keynes asserts that knowledge transcends representations. For example, he points out:

> [W]e are capable of direct knowledge about empirical entities which goes beyond a mere expression of our understanding or sensation of them. (…) [W]e are capable, that is to say, of direct synthetic knowledge about the nature of the objects of our experience. ([1921] 1973: 292–293)[3]

Keynes uses the uniformity of nature and the law of causation as examples. He also transcends the representation realm when he states:

> [Y]et in such cases [logical judgments] we believe that there may be present some element of objective validity, transcending the psychological impulsion, with which primarily we are presented. So also in the case of probability we may believe that our judgments can penetrate into the real world, even though their credentials are subjective. ([1921] 1973: 56)

The direct element does not eradicate the representative element—i.e., concept and proposition. Yet, knowledge "passes through them" towards the known thing. If knowledge stopped at the representation level, it would prove ineffective. Let me offer a golfing metaphor. A good stroke largely hinges on continuing the swing with a good "follow through." If your swing stops at the ball, the stroke proves ineffective. This "passing through"—via the representation—serves as the basis for Keynes' theory of knowledge. In Loux and Crisp's words (2017: 9),

> so far from barring us from access to things, the concepts we employ in our thinking are the vehicles for grasping the things to which they apply.

[3] "An 'analytic' sentence, such as 'Ophthalmologists are doctors', has historically been characterized as one whose truth depends upon the meanings of its constituent terms (and how they're combined) alone, as opposed to a more usual 'synthetic' sentence, such as 'Ophthalmologists are rich,' whose truth depends also upon the facts about the world that the sentence represents, e.g., that ophthalmologists are rich" (Georges Rey 2018: 1).

They are not screens or barriers between us and things; they are, on the contrary, our routes to objects, our ways of gaining access to them.

For Keynes, acquiring knowledge directly or via intuition is an act that apprehends the very nature of the thing known according to the possibilities offered by each subject matter. For example, in his *Treatise on Probability,* Keynes considers the difficulties posed by organic wholes (cf. [1921] 1973: 277, 343, 468). Keynes' anti-representationalist position stems from a philosophical reflection. Most economists share this *ethos* without any consideration. They look for the real causes of economic phenomena. Dani Rodrik, for instance, stoutly asserts, "Economics deals with the real world" (2015: 45). Economists work mostly with models, but their models aim at knowing the relevant economic aspects of reality. For example, Charles Kindleberger states (1965: 40):

An economic model is a statement of relationships among economic variables. Its purpose is to illustrate causal relations among critical variables in the real world, stripped of irrelevant complexity, for the sake of obtaining a clearer understanding of how the economy operates, and in some formulations, in order to manipulate it.

The verbs "to operate" and "to manipulate" actually refer to causes. Paul Romer's 2016 controversial paper "The Trouble with Macroeconomics" uses the verb "to cause" nine times and complains about what the author calls "post-real models" (2016: 4ff.). In turn, Keynes writes, "Economics is a science of thinking in terms of models joined to the art of choosing models which are relevant to the contemporary world" (Keynes 1973: 296, letter to Roy Harrod, 4 July 1938). He also notes, "The specialist in the manufacture of models will not be successful unless he is constantly correcting his judgment by intimate and messy acquaintance with the facts to which his model has to be applied" (Keynes 1973: 300).

In short, if we are committed to a successful economics, we are also committed to a realist metaphysics, like Aristotle's.

Metaphysical Knowledge[4]

A question remains, what is for Aristotle the instrument of metaphysics that can be used to learn about the nature of things? It is theoretical reason (from the verb *theorein*, to contemplate)—the use of human reason

[4] In this section, I took some elements from my paper 2008.

that makes it possible to know the essences (*to ti en einai*: what is) and causes (*aitiai*) that lie behind what can be empirically observed (*Metaphysics* VI, 1, 1025b 20–1 and XI, 7, 1063b 36—1064a). For Aristotle, the theoretical reason includes, first, induction (*epagogé*) or abstractive knowledge; second, intuition (*noûs*), which is a direct apprehension of the most general (first) principles of sciences (*Posterior Analytics* II, 19, 100b 5–17; *Nicomachean Ethics* VI, 6; Kahn 1995), and, finally, the deductions or scientific arguments. The theoretical reason relies on empirical data concerning physical events, but it goes beyond them. When Aristotle begins his *Metaphysics* by saying that "all men by nature desire to know" (I, 1, 980a 21), he uses the term *eidenai*—the same term that he uses to mean the knowledge of principles (*Nicomachean Ethics* VI, 7 1141a 17) and that stems from the verb *eido*, to see: it is knowing as seeing. This knowledge is not innate; it starts with the senses, the memory, and the experience, which ultimately facilitate a noetic or intuitive—not deductive—grasping of those first principles, essences, and causes. The whole process is explained in *Posterior Analytics* II, 19 (see Irwin 1990: Chapter 2, paragraphs 10–17).

Induction is part of this process, but we should not confuse Aristotle's abstractive induction (*epagogé*) with the modern concept of enumerative induction. Hintikka (1992: 34) eloquently explains the difference:

> For Aristotle, the problem of induction was not first and foremost a problem of inference from particulars to a generalization. It was a problem of concept formation. Particular cases were stepping-stones to the concepts or forms "induced" to be realized in the soul (...) Hence there is no such problem as the justification of induction for Aristotle.

These concepts or forms are already general concepts; then, we do not need to gather a complete number of instances to induce the general concept. It may be said that, for Aristotle, the theoretical truth has two levels. There is a pre-propositional level of (a richer or poorer) knowledge of essences (*Metaphysics* IX, 10, 1051b 17—1052a 4) and a propositional level of principles and judgments (*Metaphysics* IV, 7, 1011b 25–27 and VI, 4, 1027b 20–23). For Aristotle, the truth is not only a logical category associated with judgments and propositions; it refers to reality not only at the level of the actual correspondence to facts of the composition or division of terms expressed by judgments, but also at the level of the ability to grasp the actual beings that these terms designate. "Actual knowledge is identical with its objects" (*De Anima* III, 7, 431a 1): knowledge is an "intentional" possession of the form of the thing known, not a representation of it (Hintikka 2004: 46).

Theoretical reason is not intended to serve as an instrument of technique but as pure knowledge. This lack of orientation towards action is not sufficient, however, for theoretical knowledge: the essential characteristic of Aristotle's notion of theoretical reason is that it aims to know what and why things are. This does not mean that it always succeeds in its quest for true real causes. Still, it does recognize that there are real causes, and it attempts to know them.

What Is the Matter for Economics?

At this point, after making a huge effort to follow the previous pages, economist readers will surely ask this question. As I will elaborate throughout this book, economics studies the essential traits of largely non-substantial entities (properties)—like decisions or actions inhering in concrete particulars of different kinds, things or properties, individuals or institutions. In other words, actual economic actions are actions—properties—carried out by individual human beings propelled by a plethora of motives. There can be a main motive—maximization of wealth or, more recently, of utility or value—but it is a motive of a human being that has other motives that also influence her behavior.

John Stuart Mill sheds some light on the consequences of this fact. In his essay "On the Definition of Political Economy; and on the Method of Investigation Proper to It," he states[5]:

> What is now commonly understood by the term 'Political Economy' is not the science of speculative politics, but a branch of that science. It does not treat of the whole of man's nature as modified by the social state, nor of the whole conduct of man in society. It is concerned with him solely as a being who desires to possess wealth, and who is capable of judging of the comparative efficacy of means for obtaining that end. ([1844] 2006: 321)

[5] Essay V of his *Essays on Some Unsettled Questions of Political Economy* (London: Parker, 1844) published in the *London and Westminster Review* in 1836, and a second edition reprinted in 1874 with minor changes. I used the 1844 version of the Essay published in the Volume 4 of the *Collected Works of John Stuart Mill*, University of Toronto Press, 1967 (reprinted by Liberty Fund, 2006).

Mill considers Political Economy to be a limited branch of politics in much the same way as Adam Smith, who views it as "a branch of the science of a statesman or legislator" ([1776] 1828: 189). However, Mill departs from Smith because, while the latter upholds an ample conception of political economy, the former anticipates the prevailing contemporary restrictive definition of economics: the allocation of scarce means to satisfy given ends (Lionel Robbins 1935: Chapter 2).[6] Yet, Mill is also aware of the fact that this description of political economy is actually a simplifying abstraction:

> All these operations, though many of them are really the result of a plurality of motives, are considered by Political Economy as flowing solely from the desire of wealth [...] Not that any political economist was ever so absurd as to suppose that mankind are really thus constituted. ([1844] 2006: 322)

Hence, he also emphasizes the need to consider additional motives for these "operations" in order to come to a correct explanation and prediction:

> So far as it is known, or may be presumed, that the conduct of mankind in the pursuit of wealth is under the collateral influence of any other of the properties of our nature than the desire of obtaining the greatest quantity of wealth with the least labour and self-denial, the conclusions of Political Economy will so far fail of being applicable to the explanation or prediction of real events, until they are modified by a correct allowance for the degree of influence exercised by the other causes. ([1844] 2006: 323, see also 326–327)

Nancy Cartwright states that "an economic motive is something that people can have and operate from along with a number of other motives" (1994: 79). Max Weber ([1922] 1978: 24–45) distinguishes four types of motives that guide social actions: instrumentally rational, value-rational, affective, and traditional motives. According to Weber, an action is instrumentally rational when it seeks the adequate distribution of means to obtain the actor's ends. It is value-rational when it is determined by a conscious belief in the intrinsic value of some form of behavior. Affective actions are guided by the actor's affections and feelings, while traditional

[6] On Smith's "amplitude," see Dimitris Milonakis and Ben Fine (2009: 5 and 21).

actions are determined by adopted habits. Weber believes that, while a particular type of motive may prevail in some specific kind of rationality, almost all human actions stem from several of these motives.

These different motives have been considered by new, contemporary currents in economics. Amartya Sen's capability approach focuses on values; behavioral economics, on affections, feelings, and diverse psychological influences, while institutional economics zeroes in on habits. However, there is a tendency, highlighted by John Davis, to "domesticate" these new rationales of economic actions by the instrumental logic of standard economics.[7] If we consider the "accidental" condition of economic actions—human beings' actions—we discover that we must take into account the insights of these new currents' rationales.

Metaphysics can not only look at the ultimate categories of economic elements—kinds, properties—to ascertain their nature, but it can also consider their more immediate nature. For example, the nature of money—an institution, as many believe, or a social artifact?, spontaneous or designed?—helps to establish whether bitcoin and other digital moneys are indeed money. We can ask: Is the economic agent fundamentally a "maximizing" being? The answer has to do with the legitimacy of standard microeconomics theory. Why do people work? This is another relevant question. The answer to the philosophical question "Is the whole more than the sum of its parts?" contributes to clarifying the relationship between micro and macroeconomic entities.

The rest of this book will show how a metaphysical analysis of economic entities simplifies economists' work, helping them to take the right paths and to stay away from misleading ones. As noted earlier, Aristotle affirms, "he whose subject is existing things *qua* existing must be able to state the most certain principles of all things. This is the philosopher" (*Metaphysics* IV, 3 1005b 11–12).

The next chapter will introduce the metaphysical categories that will be employed to ascertain the nature of the economy and of economic entities.

[7] Davis asserts (2008: 365), "Economics, as other sciences, has regularly imported other science contents in the past, and having subsequently 'domesticated' them, remade itself still as economics. In the current situation, for example, behavioral economics—a research program in economics, not in psychology—employs imports from psychology but frames them in terms of economic concerns." In my 2017 book, I analyze whether the new currents are in fact tamed by standard economics.

References

Agazzi, E. (1988). “Science and Metaphysics: Two Kinds of Knowledge”, *Epistemologia. An Italian Journal for the Philosophy of Science* 11/11: 11–28.

Aristotle (1995). *The Complete Works of Aristotle. The Revised Oxford Translation*, J. Barnes (ed.). Princeton: Princeton University Press, 6th printing with corrections.

Batterman, R. W. (2020). “Multiscale Modeling: Explanation and Emergence”. In: W. González (ed.) *Methodological Prospects for Scientific Research. From Pragmatism to Pluralism.* Synthese Library Studies in Epistemology, Logic, Methodology, and Philosophy of Science Volume 430. Cham: Springer, pp. 53–67.

Cartwright, N. C. (1994). “Mill and Menger: Ideal Elements and Stable Tendencies”. In: B. Hamminga and N. de Marchi (eds.) *Idealization VI: Idealization in Economics.* Poznam Studies in the Philosophy of the Sciences and the Humanities 38. Atlanta and Amsterdam: Rodopi, pp. 177–188.

Cartwright, R. (1954). “Ontology and the Theory of Meaning”, *Philosophy of Science* 21: 316–325.

Crespo, R. F. (2008). Cartwright on Capacities and Sen on Capabilities and the Firm. 1st IESE Conference, “Humanizing the Firm & Management Profession”, Barcelona, IESE Business School, June 30–July 2, 2008. Available at SSRN: https://ssrn.com/abstract=1295192 or https://doi.org/10.2139/ssrn.1295192. Accessed 20 July 2020.

Crespo, R. F. (2013). *Theoretical and Practical Reason in Economics. Capacities and Capabilities.* Dordrecht: Springer.

Crespo, R. F. (2017). *Economics and Other Disciplines. Assessing New Economic Currents.* London: Routledge.

Davis, J. B. (2008). “The Turn in Recent Economics and the Return of Orthodoxy,” *Cambridge Journal of Economics* 32: 349–366.

Gilson, E. and T. Langan (1963). *A History of Philosophy. Modern Philosophy. Descartes to Kant.* New York: Random House.

Goudzwaard, B. and H. de Lange (1995). *Beyond Poverty and Affluence: Toward an Economy of Care.* Grand Rapids and Geneva: Eerdmans and WCC Publications.

Guyer, P. and A. Wood (1999). “Introduction to the *Critique of Pure Reason*”. In: Immanuel Kant, *Critique of Pure Reason*, translated and edited by Paul Guyer and Allen W. Wood. Cambridge: Cambridge University Press, pp. 1-80.

Hintikka, J. (1992). “The Concept of Induction in the Light of the Interrogative Approach to Inquiry”. In: J. Earman (ed.) *Inference, Explanation, and Other Frustrations: Essays in the Philosophy of Science.* Berkeley: University of California Press, pp. 23–43.

Hintikka, J. (2004). *Analyses of Aristotle. Selected Papers 6.* Hingham: Kluwer.

Hylton, P. and G. Kemp, "Willard Van Orman Quine", *The Stanford Encyclopedia of Philosophy* (Spring 2020 Edition), Edward N. Zalta (ed.). https://plato.stanford.edu/archives/spr2020/entries/quine/.

Irwin, T. W. (1990). *Aristotle's First Principles.* Oxford: Clarendon Press.

Kahn, C. H. (1995). "Aristotle on Thinking". In: Martha Nussbaum (ed.) *Essays on Aristotle's De Anima.* Oxford: Oxford University Press, pp. 359–380.

Kant, I. ([1787] 1999). *Critique of Pure Reason*, translated and edited by Paul Guyer and Allen Wood. Cambridge: Cambridge University Press.

Keynes, J. M. ([1921] 1973). *A Treatise on Probability, the Collected Writings of John Maynard Keynes*, Volume VIII. London: Macmillan.

Keynes, J. M. (1972). *Essays on Biography: The Collected Writings of John Maynard Keynes*, Volume X. London: Macmillan.

Keynes, J. M. (1973). *The General Theory and After: Part II. Defence and Development, the Collected Writings of John Maynard Keynes*, Volume XIV. London: Macmillan.

Kindleberger, C. P. (1965). *Economic Development*, second edition. New York: McGraw-Hill Book Company.

Loux, M. J. and T. M. Crisp (2017). *Metaphysics. A Contemporary Introduction*, fourth edition. London and New York: Routledge.

Marshall, A. ([1920] 1962). *Principles of Economics.* London: Macmillan.

Menger C. ([1883] 1985). *Investigations into the Method of the Social Sciences with Special Reference to Economics*, edited by Louis Schneider and translated by Francis Nock. New York: New York University Press.

Mill, J. S. ([1844] 2006). *Essays on Some Unsettled Questions of Political Economy* (Essay V: 'On the Definition of Political Economy; and on the Method of Investigation Proper to It'). In *Collected Works of John Stuart Mill*, Volume 4. Indianapolis: Liberty Fund.

Milonakis, D. and B. Fine (2009). *From Political Economy to Economics.* London: Routledge.

Nietzsche, F. ([1901] 1968). *The Will to Power*, translated by Walter Kaufmann and R. J. Hollingdale, edited by Walter Kaufmann. London: Weidenfeld & Nicolson.

Price, H. (1997). "Carnap, Quine and the Fate of Metaphysics", *Electronic Journal of Analytic Philosophy* 5/1, EJAP 5:1: Price, "Carnap, Quine and the Fate of Metaphysics" (louisiana.edu). Accessed 28 September 2020.

Quine, W. V. O. (1948). "On What There Is", *Review of Metaphysics* 2: 21–38.

Quine, W. V. O. (1951). "Two Dogmas of Empiricism", *The Philosophical Review* 60/1: 20–43.

Quine, W. V. O. (1953). *From a Logical Point of View.* Cambridge, MA: Harvard University Press.

Quine, W. V. O. (1966). "On Carnap's Views on Ontology". In: *The Ways of Paradox and Other Essays.* New York: Random House.

Rey, G. (2018). "The Analytic/Synthetic Distinction", *The Stanford Encyclopedia of Philosophy* (Fall 2018 Edition), Edward N. Zalta (ed.), https://plato.stanford.edu/archives/fall2018/entries/analytic-synthetic/. Accessed 29 August 2020.

Robbins, L. C. (1935). *Essay on the Nature and Significance of Economic Science*, second edition. London: Macmillan.

Rodrik, D. (2015). *Economic Rules*. Oxford: Oxford University Press.

Romer, P. (2016). "The Trouble with Macroeconomics", Delivered January 5, 2016 as the Commons Memorial Lecture of the Omicron Delta Epsilon Society. https://paulromer.net/wp-content/uploads/2016/09/WP-Trouble.pdf. Accessed 18 November 2019.

Smith, A. ([1776] 1828). *Inquiry into the Nature and Causes of the Wealth of Nations*, Book IV, Introduction, J. R. McColluch. Edinburgh, Printed for A. Black and W. Tait, 1828.

Studtmann, P. (2007). "Aristotle's Categories", *The Stanford Encyclopedia of Philosophy* (Fall 2008 edition). Edward N. Zalta (ed.), https://plato.stanford.edu/archives/fall2008/entries/aristotle-categories/. Accessed 23 October 2020.

Tahko, T. (2011). "In Defense of Aristotelian Metaphysics". In: Tuomas E. Tahko (ed.) *Contemporary Aristotelian Metaphysics*. Cambridge: Cambridge University Press, pp. 26–43.

Thomasson, A. (2018). "Categories", *The Stanford Encyclopedia of Philosophy* (Summer 2019 Edition), Edward N. Zalta (ed.), https://plato.stanford.edu/archives/sum2019/entries/categories/. Accessed 23 October 2020.

Varzi, A. C. (2011). "On Doing Ontology without Metaphysics", *Philosophical Perspectives* 25: 407–423.

Veatch, H. B. (1978). "Is Quine a Metaphysician?" *The Review of Metaphysics* 31/3: 406–430.

Weber, M. ([1922] 1978). *Economy and Society*, edited by G. Roth and C. Wittich. Berkeley and Los Angeles: University of California Press (*Wirschaft und Gesellschaft*, 1922).

Whitehead, A. N. (1929). *The Function of Reason*. Boston: Beacon Press.

CHAPTER 3

Metaphysical Categories

The last chapter has established that the metaphysical theory that will be adopted in this book is the Aristotelian metaphysics, prioritizing reality over knowledge. Economic entities will not be reduced to or defined as the subject matter of economics—quite the opposite: economics will adopt the metaphysically defined economic entities as its subject matter. The metaphysical definition of economic entities subsequently requires the definition of the categories of being that will be used in the former definition. This chapter will first introduce the Aristotelian categories.

However, the claim to undertake an ontological or metaphysical analysis of the economy is not new. This chapter will also explore three approaches setting forth interesting metaphysical categories to address the economy posited by Nancy Cartwright, Tony Lawson, and Uskali Mäki (in alphabetical order). The aim of this chapter is not to present all the ontological insights of these proposals, but to glean the categories that they employ to consider their potential use in this book. These authors take a critical look at contemporary standard economics. It is not the purpose of this chapter to focus on their criticisms, but only to pick up the concepts that might help to build an ontological analysis. Given that, as noted in the previous chapter, this book will rely on Aristotle's metaphysical theory, it will be relevant to determine whether the metaphysical theories and categories proposed by Cartwright, Lawson, and Mäki fit with the Aristotelian approach.

R. F. Crespo, *The Nature of the Economy*,
https://doi.org/10.1007/978-3-031-02453-5_3

The Aristotelian Categories of Being

Returning to Varzi's first quotation at the beginning of the last chapter, "Metaphysics seeks to explain, of those entities [existent entities], what they are" (2011: 407). Nonetheless, metaphysics does not look for the "immediate" answer—e.g., "dogs are animals." It is possible to draw a more general definition like "dogs are entities" or "dogs are things," following up with a new question: "what kind of entities or things are dogs?" For example, dogs are different kinds of things than dogs' colors. Loux and Crisp explain that "To provide a complete metaphysical theory is to provide a complete catalogue of the categories under which things fall and to identify the sorts of relations that obtain among those categories" (2017: 15). Dogs and their colors are entities, but they are different kinds of entities.

Aristotle often uses the notion of "homonymy *pròs hén*." Homonymous *pròs hén* concepts have different but related meanings. One meaning is the "focal" or primary meaning to which the others refer and are connected. Aristotle uses "healthy" as an example: the focal meaning refers to a healthy human body, while derivative meanings include healthy foods, sports, walks, medicines, and so on (cf. *Metaphysics*, IV, 2, 1003a 32 and ff.). Aristotle applies this notion to being. This is why, for Aristotle, "*to on legetai polachos*": "There are many senses in which a thing may be said to 'be', but all that 'is' is related to one central point" (*Metaphysics* IV, 2, 1003a 34–35). That is, there are different but related kinds of being, which brings us to the realm of epistemology, after laying out a provisional list of kinds.

There are a lot of "things" or "concrete particulars" in the world. Take a dog, for example—a specific dog, Max. Max has been trained as an aggressive dog. It often fights other dogs, and, as a result, it has some injuries; it is also growing old. Max might not be called Max, and it could have been trained differently, thus avoiding its fights and wounds, and it would still be the same dog all its life. What kind or sort of thing is Max, regardless of all these contingencies that happen to him? As David Wiggins puts it, "For each thing that satisfies a predicate [as in our case, fighting, training, etc.] there must exist some known or unknown, named or nameable, kind to which the items belongs and by reference to which the 'what it is' question could be answered" (2001: 21). The answer is: "Max is a dog," because "everything that exists is a *this such*" (Wiggins 2001: 22). This "such" is the kind Max is—a dog. Max would not exist if it were not a dog; Max needs to be a dog to exist. Instead, its specific properties might have been different, and these properties—its name, aggressive character, and injuries—would not exist without Max.

Aristotle considers that this entity called Max, with its specific properties, is a "substance" (*ousia*) belonging to the *universal* kind "dog." Max is an "instantiation" of the universal kind "dog." Kind is the *essence* of Max, and Max's properties are its accidents (*symbebekoi*, *accidens* in Latin, "what happens to"). The essence is necessary, and the accidents are contingent, but this dog, Max, this existent substance is not without its accidents, which—though not necessary—are part of Max's identity. Dog is not the only kind present in Max; its properties also "belong" to kinds: kinds of accidents or properties—for example, aggressiveness. Some kinds, like dogs, are always essential, while others, like aggressiveness, are always accidents or properties.

What Wiggins is pointing to was systematically developed by Aristotle. I will briefly show how.[1] First, Aristotle thinks that universal kinds only exist in concrete instantiations of kinds. Plato, his mentor, thought that kinds exist separated from the individuals that belong to them. Sophists—Socrates's contemporaries—thought that universals are only names that we assign to similar things, but they do not have a real existence, even in these things. For Aristotle, the only existent beings are individuals. However, individuals have an *eidos* or essence (a "*what* they are"), a kind or sort that belongs to all the individuals in the same *species*. The universal kind is the logical expression of the *eidos*, the "what it is," later called essence: it neither subtracts nor adds anything to the essence—it is identical to its essence. Additionally, the universal has a logical existence as the thought that contains it, and this thought has an ontological existence *qua* thought (*Metaphysics*, VII, 4, 1030a 25–27).

As Ignazio Angelelli explains (1991: 12), in his book *Categories* (Chapter V), Aristotle introduces four classes of entities (*onta*) produced by the combination of two relations. The first relation—"found in" or "being in"—unfolds between accidents and substances, and is called "inherence:" accidents "inhere" in substances. This characterization might lead to wrongfully thinking of accidents as something previously existent that "lands" on substances. Accidents exist in substances and are

[1] For a broader explanation, see my article (2006).

"part" of them. Max may be white, brown, or dark, but it needs to have a color.

The second relation—"asserted of" or "being said of"—occurs between universals and particulars. It features four classes: (1) universal substances (e.g., dog) which only exist instantiated in particulars, (2) individual substances (e.g., this dog, Max), (3) universal accidents (e.g., aggressiveness) which also only exist instantiated in particulars, and (4) particular or individual accidents (e.g., Max's aggressiveness). We can plot these classes in a diagram traditionally called the "ontological square" and adopted by Jonathan Lowe (2006). This author distinguishes between *substantial* and *non-substantial* entities, and between *particulars* and *universals*. He refers to substantial particulars as "objects," non-substantial particulars as "modes" or "tropes," non-substantial universals as "attributes," and substantial universals as "kinds." Objects are characterized by modes, which are instances of attributes. Attributes characterize kinds with objects as their instances.[2]

As David Ross explains, "In doing this, he [Aristotle] arrived at the earliest known classification of the main types of entity involved in the structure of reality" (1923: 23). This classification also surfaces in *Metaphysics* IV, 2, V, 7, VI, 4, and VII, 1. There are as many predicates as manners of existence.

The category "substance" is the focal meaning or "starting point" (*Metaphysics* IV, 2 1003b 6) of beings. Substances are, by definition, ontologically primary items: their existence can be established without relying on the existence of anything else. Substances are individual (a *tode ti*—a this), and there are identity criteria to identify every substance (cf. *Metaphysics* V, 8). The other entities are accidents. Aristotle distinguishes two types of accidents:

> We call an accident that which attaches to something and can be truly asserted, but neither of necessity nor usually, e.g., if one in digging for a hole for a plant found treasure (...) 'Accident' has also another meaning, i.e., what attaches to each thing in virtue of itself but is not in its substance, as having its angles equal to two right angles attaches to the triangle. And accidents of this sort may be eternal, but no accident of the other sort is. (*Metaphysics* V, 30, 1025a 30–34)

[2] There is a discussion between Aristotelian specialists about Aristotle's ontological conception of these classes. I will not engage in it.

The first type is casual, not necessary. The second type is what, while being a property, necessarily pertains to the substance or accident in which it inheres—for example, for material bodies to have an extension (accident). This second type of accidents are essential properties or accidents, called "*proprios*" by Medieval Scholastic thinkers. An accident happens to a substance either immediately or in a mediated way (through another accident). Aristotle provides examples of substance and accidents:

> To give a rough idea, examples of substance are man, horse; of quantity: four-foot, five-foot; of quality: white, literate; of a relative: double, half, larger; of where: in the Lyceum, in the market-place; of when: yesterday, last year; of being-in-a-position: is-lying, is-sitting; of having: has-shoes-on, has-armour-on; of doing: cutting, burning; of undergoing: being-cut, being-burned. (*Categories* 4, 1b27–2a4)

"Relations" are specific bridges between substances: they are "binomial" properties. Relations may involve quantity, like larger or greater, but also quality, as fatherhood.

A final question, what is the ontological condition of artifacts, like a bed, a table, or a clock? Aristotle speaks of them when referring to nature. For him, "Nature is a source or cause of being moved and of being at rest in that to which it belongs primarily, in virtue of itself and not in virtue of a concomitant attribute" (*Physics* II, 1, 192b 22–24). Nature implies a natural subject that is a substance (*Physics* II, 1, 192b 32–35): "animals and their parts, plants, and simple bodies like earth, fire, air and water" (*Physics* II, 1, 192b 10–11), and obviously, man.

Concerning artifacts, Aristotle explains, "A bed and a coat and anything else of that sort, qua receiving these designations, i.e., in so far as they are product of art, have no innate impulse to change. But in so far as they happen to be composed of stone or of earth or of a mixture of the two, they have such an impulse" (*Physics* II, 1, 192b 16–22). The internal adequate disposition of the parts of artifacts allows them to operate.

I will use the words "things," "objects," "concrete particulars," and "substances" indistinctively, as well as the words "accidents," "attributes," and "properties," and the words "kind," "essence," and "nature." Still, I will prefer the more modern choices—for example, properties rather than accidents.

Nancy Cartwright's Categories[3]

I have stated that we need to develop a metaphysical analysis of the economy and its components, which requires a metaphysical theory that allows for a knowledge going beyond merely "saving the phenomena." This last expression means delivering unified descriptions of natural regularities among things compatible with the observable, without trying to delve into unobservable underlying entities. Nonetheless, in order to save regular phenomena, it is necessary to commit to causal mechanisms that can be gleaned from data, but that are not directly recorded by human perceptual systems or experimental equipment (see Bogen 2009). For James Bogen and James Woodward (see Bogen and Woodward 1988; Woodward 1989: 393), phenomena are stable and general features of the world that go beyond data, and that can be explained and predicted by general theories. Theories, for these authors, are not about data, but about phenomena. Phenomena, explains Bogen (2009), are processes, causal factors, effects, facts, regularities, and other pieces of ontological furniture. This implies that knowledge goes beyond observation; observations only help us to get to the knowledge of those kinds of phenomena.

Agreeing with Bogen and Woodward (1988) and quoting them, Cartwright (1989: 169) argues that "nature is full, not only of data, but of phenomena as well." She understands scientific explanations in terms of stable causes which she calls "capacities" or "natures" (Cartwright 1992: 71, nt. 7). This claim has Aristotelian roots, which she herself acknowledges. For her, those phenomena considered by Bogen and Woodward include capacities and interactions. Her general program aims at defining what capacities are (ontology), how they are construed (epistemology), and how we use them (Cartwright 2007: 1). She also attempts—somewhat skeptically—to apply this notion of scientific explanation to the social realm—specifically to economics.

Cartwright opposes Hume's reduction of causality to the mere regularity of association: "The generic causal claims of science are not reports of regularities but rather ascriptions of capacities, capacities to make things happen, case by case" (Cartwright 1989: 2–3). Cartwright's following quotation (1989: 211) is clear:

[3] This section draws from some passages of Crespo (2013, Chapter 2).

> I chose deliberately the Aristotelian language of matter, form, and function because these terms are fundamental to a preliminary description of phenomena that appear in my image of science. This language is a threat to the neo-Humean covering-law theorist, and it is meant as such.

Matter and form are considered by Aristotle as "co-principles" of substance—a doctrine called "hylomorphism."[4] In *Physics*, Aristotle deals with the movements and changes taking place in substances: accidental—i.e., qualitative, quantitative, local, inhering in the substance—and the generation or corruption of the very substance. In reference to this last type, he explains:

> If, then, we grant that the things of Nature have ultimate determinants and principles which constitute them, and also that we can speak of them 'coming to be' not in an incidental but in an essential sense—[...]—then it is obvious that they are composed, in every case, of the underlying subject and the 'form' which their defining properties give to it. (*Physics* I, 7, 190b 18–21)

The underlying subject of substantial changes is *próte hýle* or prime matter (*Physics* II, 1, 193a 29). For Aristotle, this composition does not imply that matter and form are independent parts of the substance: they "cannot even exist if severed from the whole" (*Metaphysics* VII, 10, 1035b 24–25). He also affirms, "The final matter and the form are one and the same thing, the one potentially, and the other actually" (*Metaphysics* VIII, 6, 1045b 18–19). The form is the actuality and unifier of being, not an element but a principle (cf. *Metaphysics* VII, 17, 1041b 25–31). Anna Marmodoro explains, "The substance is a composite of matter and form, and yet *one*" (2013: 20). Concerning "function," Aristotle views it as a way of functioning of the substance. This way depends on the nature and end of a specific substance. The human function, for example, is a rational way of doing human activities (*Nicomachean Ethics* I, 7).

In addition, Cartwright agrees with John Stuart Mill's proposal about the existence of "tendencies," which she correlates with "capacities": "I suggest that the reader take my 'capacity' and Mill's 'tendency' to be synonymous" (Cartwright 1989: 170). According to Cartwright, Mill's

[4] Hylomorphism is currently a very active doctrine with many scholars working on it. For a review of the main discussions, authors and bibliography, see Anna Marmodoro and Michele Paolini Paoletti (2021).

tendencies are not tendencies of events but tendency factors or stable real causes. For Cartwright, Mill's idea of tendency pertains to "the essential behaviour of a factor" (1989: 203).

In Book III, Chapter X of his *System of Logic*, "Of the Plurality of Causes, and of the Intermixture of Effects," Mill argues that one phenomenon can be produced by different causes: "It is not true, then, that one effect must be connected with only one cause, or assemblage of conditions" (1882: 311). One phenomenon may involve a concurrence of causes, which may happen in two different ways. First, the different causes modify or interfere with each other's effects, thus constituting a compound causal action. Mill uses the joint operation of different forces in mechanics to exemplify this. Second, "Illustrated by the case of chemical action, the separate effects cease entirely, and are succeeded by phenomena altogether different, and governed by different laws" (1882: 315). In the first case, Mill explains the action of each cause by saying that "it *tends* to move in that manner even when counteracted" (1882: 319: italics in the original). He concludes, "All laws of causation, in consequence of their liability to be counteracted, require to be stated in words affirmative of tendencies only, and not of actual results" (1882: 319). From this, Cartwright (1882: 179) concludes that "Mill's view has to be that the fundamental laws of nature are laws that assign stable tendencies to specific causes," which is Cartwright's notion of capacities.

However, Cartwright has been criticized for her interpretation of Mill: her concept of capacity differs from Mill's notion of tendency. Christoph Schmidt-Petri (2008) argues that Cartwright's capacities are significantly different from Mill's tendencies, which he also believes to be problematic for Mill's entire thinking. According to Schmidt-Petri, Mill uses the concept of tendency for entirely practical methodological reasons rather than for metaphysical reasons (2008: 292). Hence, Mill's tendency notion is not really compatible with Cartwright's realistic view of capacities (2008: 298). In her reply to Schmidt-Petri (2008), she has admitted that she was possibly wrong in applying her concept of capacity to Mill. In sum, the difference between Cartwright's capacities and Mill's tendencies is that, while for her capacities are clearly and always real stable causes, for Mill the concept of tendency is only a methodological device that does not necessarily express an ontological reality. That is, Cartwright's conception of capacity implies a metaphysical commitment.

Thus, Cartwright proves to be more Aristotelian than Millian. In Chapter 6 of *The Dappled World* (reprinted with slight changes in 2001), she explains:

> [The thesis that] I am most prepared to defend follows Aristotle in seeing natures as primary and behaviours, even very regular behaviours, as derivative. Regular behaviour derives from the repeated triggering of determinate systems whose natures stay fixed long enough to manifest themselves in the resulting regularity. (1999: 149; 2001: 290)

In Chapter 3 of the same book, she asks, "What facts then are they that make our capacity claims true?" She concludes:

> [T]he best worked out account that suits our needs more closely is Aristotle's doctrine on natures, which I shall defend in the next chapter. Capacity claims, about charge, say, are made true by facts about what it is in the nature of an object to do by virtue of being charged. To take this stance of course is to make a radical departure from the usual empiricist view about what kind of facts there are. (1999: 72)

On "Aristotelian Natures and the Modern Experimental Method" (1992), Cartwright persuasively shows that what science actually does by studying "the inner constitution [of things and events] is a study of an Aristotelian-style nature" (1992: 69):

> Still, I maintain, the use of Aristotelian-style natures is central to the modern explanatory program. We, like Aristotle, are looking for 'a cause and principle of change and stasis in the thing in which it primarily subsists' [*Physics* II, 1, 192b22], and we, too, assume that this principle will be 'in this thing of itself and not *per accidens*'. (1992: 47)

According to Cartwright, capacities, natures, or "powers to do" are real causes (cf., e.g., 1989: 182). They feature three elements: (1) potentiality: what a factor can or tends to do in the abstract; (2) causality: they are not mere claims about co-association; and (3) stability (Cartwright 1998: 45). She calls them "natures" (1992), while capacities are internal forces, or "inner causes."

For Aristotle, a capacity or *dynamis* is a "power to do." He defines it as "a source of movement or change, which is in another thing that the thing moved or in the same thing *qua* other" (*Metaphysics* V, 12,

1019a 15–16). *Dynamis* is an "urge of nature to grow to maturity, to realize form, and to perform the due function" (Guthrie 1967: 140).[5] Concerning causes, Aristotle uses the idea of potentiality in reference to material cause. However, for both Cartwright and Aristotle, the causal structure of a nature (formal cause) is the most relevant cause in the very being and in the scientific explanation of a concrete phenomenon. Causes, in any case, are the four kinds of causes considered by Aristotle—material and formal, efficient, and final (*Metaphysics* I, 3–10; *Physics* II, 3)—that allow different types of explanations, "a doctrine of four becauses" (Ackrill 1981: 36) that answers to these questions: Of what is this made? (material cause), why is it this thing and not another? (formal cause), who made it? (efficient cause), and for what is this made? (final cause).

When she refers to capacities' stability and applicability (1989: 146; see also 1992: 51), Cartwright states that "capacities are much like essences." In this regard, she asserts that her notion of capacities has Aristotelian resonances (1992: 45–48, 69; 1999: 72; 2001: 277, 290). Among the Aristotelian causes, she assigns priority to the *form*, which is similar to the causal structure (1989: 223).

How do we recognize capacities? This is not an easy task. Cartwright maintains that stable causes or capacities are known by intellectual abstraction (1989: 8, Chapter 5). She also shows that capacities—under specific (and difficult to achieve) conditions—can be deduced from probabilities, and that they can be measured (1989: 1.4 and 2.4). Nonetheless, this relies on the assumption that we have some causes to begin with: "no causes in, no causes out" (1989: Chapter 2).

To measure capacities is not the same as understanding them. We may measure some effects or some things that cause other things, but not the causation itself. "We cannot, of course, tell by measurement itself that what we are measuring is a real capacity" (Cartwright 2007: 42, nt. 57). Here, theoretical reason is needed. Still, measuring proves crucial to have initial experimental contact with data that manifest causes and effects and thus provide abstract knowledge on them. Measures induce or enable us to infer an abstract knowledge of causation (Cartwright 2007: 178). This involves a process of subtracting the concrete circumstances and the material in which a cause is embedded and all that follows as a result of this

[5] *Dynamis* is a power, might, strength; an ability to do something, a faculty, a capacity. Source: H. G. Liddell and R. Scott, Greek-English Lexicon (Oxford, Clarendon Press, 1900).

(Cartwright 1989: 187). In conclusion, this Aristotelian analysis confirms the real and profound nature of Cartwright's capacities and the need for them to be known via theoretical reason.

Nature, capacities, matter and form, function, and the four Aristotelian causes are the metaphysical categories used by Cartwright, and they will prove useful in the metaphysical analysis undertaken in this book.

Tony Lawson's Categories

Cambridge economist Tony Lawson also calls for a metaphysical analysis of economic reality. Lawson, based on John Locke's *Essay Concerning Human Understanding*, views philosophers as under-laborers "clearing ground a little and removing some of the rubbish that lies in the way to knowledge" (Locke [1690] 1985, xlii–xliii). Lawson starts by pointing out his views on the "rubbish" of standard contemporary economics. I am not concerned with that here. I find it interesting that he thinks "ontology analysis constitutes just the sort of philosophical under-labouring which economics at this time most needs" (2004: 324). That is, for Lawson, ontology comes before science: it is an indispensable pre-scientific task. We need to know the nature of a science's subject to establish the appropriate method to learn about it. For Lawson, ontology is an "enquiry into (or a theory of) the nature of being or existence" (2004: 324) and its role is the

> elaboration of as complete and encompassing as possible a conception of the broad nature and structure (of a relevant domain) of reality as appears feasible. The aim is to derive a general conception that seems to include all actual developments as special configurations. Put differently, a central objective is to provide a categorial grammar for expressing all the particular types of realisation in specific contexts. (2003: xvi)

For Lawson, as for Cartwright, we need to recognize the structures, powers, mechanisms, and tendencies that are irreducible to but underpin the actual course of events and states of affairs. Stephen Pratten (2007: 240–241), in an article showing the strong similarities between Cartwright's and Lawson's ontological categories, describes the main characteristics of Lawson's proposed ontology (and see Lawson 1997: 21ff.):

> Central to it though is the idea that reality is structured in the sense of not being exhausted by events and states of affairs. Consider an aspirin. In virtue of its intrinsic chemical structure it has certain powers, such as to relieve pain or thin the blood. Or consider a bicycle. Because of its physical structure it facilitates rides. Now the powers of aspirins, bicycles and anything else, can exist unexercised. When powers are exercised they work by way of mechanisms or processes. [...] The category of tendency is reserved to capture the idea that something can be continually active even if its effect is not completely actualised in the outcome. If I have a headache, the aspirin will have a tendency to make it better even as other tendencies intervene so as to ensure that the headache intensifies. It should be emphasised that the operation of a tendency in an open system, i.e., in the face of countervailing tendencies, licenses claims that are transfactual as opposed to being merely counterfactual. They do not inform us of what would happen if things were different but of what is happening in reality whatever the actual outcome. They tell us for instance that the gravitational tendency acts on the cup in my hand whatever I do with it. The gravitational tendency is not merely something that would, counterfactually, have an impact if the cup were dropped in an experimentally constructed vacuum; it is something that is acting on the cup and continues to do so whether I drop it in a vacuum, juggle with it or rest it on a table. [...] For laws of nature are not considered to express event regularities, but rather the workings of underlying mechanisms and tendencies.

Lawson draws extensively from philosopher Roy Bhaskar's work on "transcendental realism." For Lawson, following Bhaskar, reality is open and stratified. First, it is open as opposed to closed. Closed systems are those that always necessarily produce the same regularities. By contrast, Lawson affirms that we live in "a world that is complexly structured, open, intrinsically dynamic, characterised by emergence and so novelty, and inclusive of totalities and causally efficacious absences, amongst other things" (1997: 65). He asserts that these characteristics specially apply to the social realm, including the economy.

Second, reality is stratified: "the real (including the necessary and the possible) is irreducible to the domain of the actual (instances of the possible, the actual course of events) which in turn is irreducible to the empirical or conceptual" (1997: 62). The real is the stage of the underlying structures, powers, mechanisms, and tendencies, while the actual is the stage of the actual events and states of affairs, which can be captured at the empirical or conceptual level. There are underlying causes of events and phenomena: "the basic view is of a world composed in part of complex things and situations which, by virtue of their *structures*, possess certain *powers*—potentials, capacities, or abilities to act in certain ways […] A mechanism is basically a way of acting of a structured thing" (2003: 144). Tendencies are the characteristic ways of acting that mechanisms have but are not always actualized.

Not without tensions, Bhaskar and Lawson seem to escape the Kantian trap. Lawson picks up Bhaskar's "epistemic fallacy" notion, reducing ontology to epistemology, giving the category of the former to the latter. He maintains that we have to look to the social realm directly (1997: 53, 57).

By contrast, it seems that they do not escape a materialist conception. Lawson explains that human agency with its intentional causality is "sustained by an emergent powers materialist orientation […] Indeed, it is clear that the form of materialism to which we are committed ultimately entails the unilateral ontological dependence of social upon biological upon physical forms coupled with the taxonomic and causal irreducibility of each to any other" (1997: 63). There are "real material causes or structures which facilitate intentional action" (1997: 31 and see 58). There is a real tension in this stance because, at the same time, Lawson argues that the irreducibility of the strati allows for the defense of freedom of human choice (see 1997: 9, 39, 174, 175, 186). He criticizes the determinism of standard economics "instead of transformative intentional agency" (1997: 65).

Regardless of these tensions, Lawson proposes some metaphysical categories—structures, powers, mechanisms, and tendencies—that will also prove useful in the analyses presented in this book.

Uskali Mäki's Categories

Finnish professor of philosophy of economics Uskali Mäki's key message is that we must undertake the mission set forth in this book: to develop an economic ontology. He states, "I believe the study of economic ontology is a prerequisite for understanding economics as a scientific discipline" (2001b: xv). For him, the discontinuity between empirical evidence and theory underscored by the Duhem-Quine thesis can only be overcome by ontology, which can justify methods and theories (cf. 2001a: 9).

Thus, Mäki construes ontology in the classical sense, as the study of being as being (2001b: 7). Thus, Mäki's questions are: "What is the economy made of? What are its constituents, and how do they hang together? [...] The study of economic ontology is concerned with what may be called 'the economic realm': the economic realm consists of those parts or aspects of the universe which are set apart as constituting the subject matter of economics" (Mäki 2001a: 3 and 4). Mäki calls this research a "local or regional" ontology (Mäki 2001a: 7).

Mäki is an ontological realist: he thinks that there is an external reality. He identifies different types of realism (see Mäki 1989, 1998b), distinguishing between realism—a philosophical meta-theory, i.e., a theory of theories—and realisticness—a property of a theory or representation. Mäki suggests the expression that something "exists recognition independently" as appropriate to completely avoid circumscribing the scope of ontological realism to the physical realm (Mäki 1998b: 406).

Ontological realism constitutes the foundation of the other kinds of realism. First, logical-semantic realism states that the propositions about entities—with respect to which there is an ontological commitment—are true (or false) if the conditions for truth of these propositions hold (or not) determinatively, objectively, and independently of our knowledge capacities. As Mäki says, "(...) semantic realism is the thesis that the theses contained in scientific theories are genuine, true or false, statements about the real world and that they have a truth value irrespective of whether we are able to determine it" (Mäki 1998b: 406). For logical-semantic realists, there are objective criteria of truth. This is the realism that is connected to a theory of truth.

Finally, Mäki describes epistemic realism as follows:

> the Xs that are claimed to exist are also knowable. Different forms of epistemological realism presuppose some versions of ontological realism and semantic realism and add to them the idea of being known or being knowable. Epistemological realism says of some existing X that facts about X are known or can be known, implying that knowers have epistemic access to X, that there is no veil separating the cognitive subject and the existing object. (Mäki 1998b: 407)

Referring specifically to the economy, Mäki often asserts that it is composed of commonsense frameworks and experiences (e.g., 1998a: 306, 2000: 111). He posits that "ontologically speaking, economics appear to be closely linked to common-sense conceptions about the world" (1998a: 301). Also, economic entities are mind-dependent (1992: 431–433; 1998a: 303, 311), but they "can exist externally even though their existence is causally dependent on the human mind" (1990: 294). He adds that these entities exist "inquiry-independently" (220: 93).

Mäki's perspective "goes" from economics to the economy. In other words, his goal is to ascertain the ontological nature of the entities addressed by economics. This means that the entities he proposes to investigate are defined by the science—as quoted earlier, "the economic realm consists of those parts or aspects of the universe which are set apart as constituting the subject matter of economics" (Mäki 2001a: 4). This direction does not seem to be the classical approach, but rather the modern one. However, his notion that economics deals with commonsensibles is an indirect way of drawing away from the modern direction. As Mäki himself asserts, "there is not departure from the world occupied by commonsensibles in economics (that is, in 'scientific economics' in contrast to 'folk economics')" (2000: 111).[6] Economics works with these commonsensibles (2002: 95). Additionally, Mäki contributes with a rich definition of categories that will be useful in the analysis of this book. Which categories has he established?

Mäki is an analytic thinker who loves definitions and subtle distinctions. Borrowing the expression from Coase and Richardson, he claims that "the way the world works" imposes a constraint on theorizing (1998a: 314; 2001c: 371, 377). He presents the pairs "primary and secondary factors, essential and accidental properties, key features and incidental features, major and minor causes, causally and less relevant factors" (1998a: 312): models should contain the first component of each pair. For Mäki, abstractions or assumptions cannot do away with these elements. We should know what is essential and what is not essential (1998a: 312; 2001c: 377–379). He asserts, "An isolating theory or statement is true if it correctly represents the isolated essence of the object; otherwise it is false" (1992: 344).

Mäki recognizes that both Nancy Cartwright and himself have been influenced by the Aristotelian tradition (2008: 4). He also notes that there are many similarities between Cartwright's (1989) account of economic theory and his own: they both believe in capacities and causal powers (2009: 14). He also speaks about causal processes and interactions (2001c: 371) as well as mechanisms (1998a: 315, 317). He also mentions entities and properties (2001a: 4), things, complex of properties, structures, and processes (2002: 92). For Mäki, "being real by no means implies being observable" (2002: 95).

[6] Margaret Schabas is skeptical about this continuity. She affirms, "Whatever is meant by 'The Economy' in folk economics appears to be significantly divergent form what is posited in scientific economics" (2009: 3). I will not delve into this difficult topic here.

Conclusion

This chapter has introduced the set of metaphysical categories that will be employed to establish the nature of the economy and its components—things or objects, attributes or properties, kinds or essences or natures, forms or structures, capacities, stable causes or powers, mechanisms and tendencies, processes. These categories are Aristotelian, expressed in a classical or a modern way. The next chapter will present the analysis of the nature of the economy using these categories coined by Aristotle, Cartwright, Lawson, and Mäki. The subsequent chapters—on the economic agent, macroeconomic entities, and money—will also define the nature of these constituents of the economy taking these categories into account.

References

Ackrill, J. L. (1981). *Aristotle the Philosopher*. Oxford: Clarendon Press.

Angelelli, I. (1991). "Accidents III: The Ontological Square". In: B. Smith and H. Burkhardt (eds.) *Handbook of Metaphysics and Ontology*. München: Philosophia Verlag.

Aristotle. (1941). *The Basic Works of Aristotle*, edited and with an Introduction by Richard McKeon. New York: Random House (reprint of the translations prepared under the editorship of W. D. Ross. Oxford: Oxford University Press).

Aristotle. (1954). *Nicomachean Ethics*, translated and introduced by Sir David Ross. Oxford: Oxford University Press.

Aristotle. (1995). *The Complete Works of Aristotle. The Revised Oxford Translation*, J. Barnes (ed.). Princeton: Princeton University Press, 6th printing with corrections.

Bogen, J. (2009). "'Saving the Phenomena' and Saving the Phenomena". http://philsci-archive.pitt.edu/archive/00004554/01/Sumitted_'Saving'-Saving.doc. Accessed 19 January 2020.

Bogen, J. and J. Woodward (1988). "Saving the Phenomena", *The Philosophical Review* 97/3: 303–352.

Cartwright, N. C. (1989). *Nature's Capacities and their Measurement*. Oxford: Oxford University Press.

Cartwright, N. C. (1992). "Aristotelian Natures and the Modern Experimental Method". In: John Earman (ed.) *Inference, Explanation, and Other Frustrations*. Berkeley, Los Angeles and Oxford: University of California Press, pp. 44–71.

Cartwright, N. C. (1998). "Capacities". In: John Davis, D. Wade Hands and Uskali Mäki (eds.) *The Handbook of Economic Methodology*, Elgar, pp. 45–47.

Cartwright, N. C. (1999). *The Dappled World: A Study of the Boundaries of Science*. Cambridge: Cambridge University Press.

Cartwright, N. C. (2001). "*Ceteris* Paribus Laws and Socio-economic Machines". In: U. Mäki (ed.) *The Economic World View. Studies in the Ontology of Economics*. Cambridge: Cambridge University Press, pp. 275–292.

Cartwright, N. C. (2007). *Hunting Causes and Using Them: Approaches in Philosophy and Economics*. Cambridge: Cambridge University Press.

Cartwright, N. C. (2008). "Reply to Christoph Schmidt-Petri". In: S. Hartmann, C. Hoefer and L. Bovens (eds.) *Nancy Cartwright's Philosophy of Science*. London: Routledge, pp. 303-304.

Crespo, R. F. (2006). "The Ontology of the 'Economic': An Aristotelian Analysis", *Cambridge Journal of Economics* 30/5: 767–781.

Crespo, R. F. (2013). *Theoretical and Practical Reason in Economics. Capacities and Capabilities*. Springer.

Guthrie, W. K. C. (1967). *The Greek Philosophers*. London: Methuen.

Lawson, T. (1997). *Economics and Reality*. London: Routledge.

Lawson, T. (2003). *Reorienting Economics*. London: Routledge.

Lawson, T. (2004). "Philosophical Under-Labouring in the Context of Modern Economics". In: J. B. Davis, A. Marciano and J. Runde (eds.) *The Elgar Companion to Economics and Philosophy*. Cheltenham and Northampton: Elgar, pp. 317–338.

Locke, J. (1690 [1985]). *An Essay Concerning Human Understanding*. In an abridgment selected by J. Yolton (ed.) London and Melbourne: Dent.

Loux, M. J. and T. M. Crisp (2017). *Metaphysics: A Contemporary Introduction*, fourth edition. London and New York: Routledge.

Lowe, J. (2006). *The Four-Category Ontology: A Metaphysical Foundation for Natural Science*. Oxford: Oxford University Press.

Mäki, U. (1989). "On the Problem of Realism in Economics", *Ricerche Economiche* XLIII/1–2: 176–98.

Mäki, U. (1990). "Mengerian Economics in Realist Perspective", *History of Political Economy*, Annual Supplement to 22: 289–310.

Mäki, U. (1992). "On the Method of Isolation in Economics", *Poznam Studies in the Philosophy of Science and Humanities* 26: 317-351.

Mäki, U. (1998a). "Aspects of Realism About Economics", *Theoria* 13/2: 301–309.

Mäki, U. (1998b). "Realism" and "Realisticness", In: J. B. Davis, D. W. Hands, U. Mäki (eds.) *The Handbook of Economic Methodology*. Cheltenham and Northampton: Elgar, pp. 404–413.

Mäki, U. (2000). "Reclaiming Relevant Realism", *The Journal of Economic Methodology* 7: 109–125.

Mäki, U. (2001a). "Economic Ontology: What? Why? How?" In: U. Mäki (ed.) *The Economic World View: Studies in the Ontology of Economics*. Cambridge: Cambridge University Press, pp. 3–14.

Mäki, U. (2001b). "Preface". In: U. Mäki (ed.) *The Economic World View: Studies in the Ontology of Economics*. Cambridge: Cambridge University Press, pp. xv–xvi.

Mäki, U. (2001c). "The Way the World Works (www): Towards an Ontology of Theory Choice". In: Mäki (ed.) *The Economic World View: Studies in the Ontology of Economics*. Cambridge: Cambridge University Press, pp. 369–389.

Mäki, U. (2002). "Some nonreasons for Nonrealism About Economics". In: U. Mäki (ed.) *Fact and Fiction in Economics*. Cambridge: Cambridge University Press, pp. 90–104.

Mäki, U. (2008). "Philosophy of Economics". In: Stathis Psillos and Martin Curd (eds.) *The Routledge Companion of the Philosophy of Science*. London: Routledge, pp. 543–554.

Mäki, U. (2009). "Realistic Realism About Unrealistic Models". In: Harold Kincaid and Don Ross (eds.) *The Oxford Handbook of the Philosophy of Economics*. Oxford: Oxford University Press, pp. 68–98.

Marmodoro, A. (2013). "Aristotle's Hylomorphism without Reconditioning", *Philosophical Inquiry* 36/1–2: 5–22.

Marmodoro, A. and M. Paolini Paoletti (2021). "Introduction to the Special Issue on form, Structure and Hylomorphism", *Synthese* 198: 2647–2656.

Mill, J. S. (1882). *A System of Logic. Ratiocinative and Inductive: Being a Connected View of the Principles of Evidence and the Methods of Scientific Investigation*, eight edition. New York: Harper & Brothers Publishers.

Pratten, S. (2007). "The Scope of Ontological Theorising", *Foundation of Science* 12: 235–256.

Ross, D. (1923). *Aristotle*. London: Methuen.

Schabas, M. (2009). "Constructing 'The Economy'", *Philosophy of the Social Sciences* 39/1: 3–19.

Schmidt-Petri, C. (2008). "Cartwright and Mill on Tendencies and Capacities". In: S. Hartmann, C. Hoefer and L. Bovens (eds.) *Nancy Cartwright's Philosophy of Science*. London: Routledge, pp. 291–302.

Varzi, A. C. (2011). "On Doing Ontology without Metaphysics", *Philosophical Perspectives* 25: 407–423.

Wiggins, D. (2001). *Sameness and Substance Renewed*. Cambridge: Cambridge University Press.

Woodward, J. (1989). "Data and Phenomena", *Synthese* 79/3: 393–472.

CHAPTER 4

The Metaphysics and Ontology of the Economy

In Chapter 2, I anticipated that "Chapter 4 will tackle two questions—'what is the nature of the economy?' in metaphysical terms, and 'which are the economic entities?' from an ontological standpoint." Thus, this chapter will have two main sections: one on the metaphysics and the other on the ontology of the economy. The former will include several subsections and a conclusion, while the section on the ontology or inventory of economic entities will be short: most constituents of economic entities will surface while dealing with the nature of the economy.

The Metaphysics of the Economy

In *Topics*, one of the books of his *Organon* (a title coined by ancient commentators to gather Aristotle's books about ordinary and scientific reasoning and logics), Aristotle proposes a set of norms to correctly undertake dialectical discussions. One of these rules refers to the appropriate use of terms. He states, "We ought to use our terms to mean the same things as most people mean by them" (*Topics* II 2 110a 16–17). However, he adds as an example, "It is right to call 'healthy' whatever tends to produce health, as do most men (*polloí*), but in saying whether the object before us produces health or not, we should adopt the language no longer of the multitude but of the doctor" (*Topics* II 2 110a 19–22). That is, we should first refer to things in the way people do it; science adds

R. F. Crespo, *The Nature of the Economy*,
https://doi.org/10.1007/978-3-031-02453-5_4

precisions only afterward. To go for a walk is usually healthy, but in some circumstances (a stormy day) or for some people (very sick individuals), it might not prove healthy.

Then, the first step that we must take in this journey to the metaphysics of the economy is to establish its ordinary meaning. As Julio Soler Miralles pointed out years ago, a reflection on the philosophy of the economy "begins with the economy-reality or sector of the real universe that is generically called 'the economic,' which is to be subsequently a consideration of the conceptual system constituted by the human mind by cognitively penetrating the economic reality" (1952: 129–39, 133).

Establishing the meaning of the economy or "penetrating the economic reality" is not an easy task. I have discussed this with Mäki more than once. He has told me that he considers this as the most difficult topic for a philosophy of the economy, but he wrote to me, "I promise that once I have a paper on the economy, I'll share it with you!" (e-mail message dated 5 May 2018). It is a bit paradoxical that, though he believes that the study of economic ontology is a prerequisite for understanding economics (2001b: xv), he has not written about this topic yet. However, I believe him; my experience with Uskali is that he thinks and works a lot on a topic before writing about it. The same paradoxical conclusion applies to Tony Lawson: the entry "economy" does not show up in the subject indexes of his books. Nonetheless, he offers some ideas about the nature of the economy that I will introduce. For my part, I have discussed some thoughts on this topic in my 2011 paper and in Chapter 2 of my book in 2013, elaborating on them in Chapter 2 of my book in 2020. In this chapter, I will try to build on my latest works, adding the ideas of the thinkers whose metaphysical categories I introduced in the previous chapter: Aristotle, Cartwright, Lawson, and Mäki. However, the essential conclusions are similar.

The section is organized as follows. First, I will try to establish the ordinary meaning of the economy, looking for it in dictionaries and complementing its definition with some of its characteristics. Then, I will introduce the concepts of the economy developed by Aristotle, Cartwright, Lawson, and Mäki, and their metaphysical analyses of them. Finally, I will conclude summarizing all the findings.

Two Meanings of the Economy[1]

Margaret Schabas states that "Economists study 'The Economy,' or so one might suppose. Yet this overarching entity is strikingly absent from mainstream economics" (2009: 3). She also remarks that the entry "economy, the" is absent in dictionaries of economics. When talking about "the economy," she refers to the whole economic system.[2] However, as I have noted in my 2020 book (2020: 16), "'the economy' is not only the economic system. There are a lot of other things that are also 'economic': actions, decisions, goods, ideas, policies, strategies, news, ministers, sources of energy, houses, cars, and so on." That is, we do not only have the noun but also the adjective that indicates an economic condition. In addition, there are some things that are more "economical" than others, and we can act "economically," "economizing" or not. Concerning the whole economic system, I will come back to this "entity" in Chapter 8, when dealing with macroeconomy.

The meanings for "economy" found in dictionaries are numerous. I have chosen the ones I consider more relevant. "Economy" is "the management of the resources of a community," "the management of household affairs," "an efficient use of something." "Economic" pertains "to the production, distribution and use of income, wealth, and commodities"; it refers "to an economy, or system of organization." "Economical" "implies careful and saving use of resources." "Economically" is to act "as regards the efficient use of income and wealth." "To economize" is "to use sparingly or frugally" (*Webster's Encyclopaedic Unabridged Dictionary of the English Language*, 1996: 618).[3]

A primary conclusion of this listing of meanings is that "the economic" is an analogous expression—i.e., it has several meanings that are somewhat

[1] I draw from some parts of my article 2011.

[2] This should not come as a surprise because, as I explained in Chapter 2, this definition corresponds to the philosopher; it is a task of the philosophy of economics.

[3] I also took into account some definitions of the Cambridge Dictionary, Economy: "the system of making money and producing and distributing goods and services within a country or region," "the system of trade and industry by which the wealth of a country is made and used," "the careful use and management of money or of time, energy, words, etc." Economic: "relating to trade, industry, and money," "making a profit, or likely to make a profit"; Economical: "not using a lot of money"; Economically: "using little money, time, etc."; Economize: "to try to save money by reducing the amount that you are spending." (https://dictionary.cambridge.org/es/diccionario/ingles/).

alike and somewhat different. Therefore, we must determine the analogous meanings of "the economic" and how they interrelate with each other.

Several classifications of "analogy" have been proposed. A widely accepted one distinguishes between the "analogy of attribution" and the "analogy of proportionality." Things are analogous by attribution when there is a primary meaning, separate from other meanings derived from the former. Analogy of attribution corresponds to Aristotle's concept of homonymy *pròs hén*, explained in the previous chapter. Homonymous *pròs hén* concepts have different related meanings, one of which is the "focal" or primary meaning to which the other derivative meanings refer and are connected. In the previous chapter, I used the example of health and applied it to being—with substance as the focal or primary analogue and accidents as derivative analogues. Analogy of attribution can be intrinsic or extrinsic, depending on whether it is an analogous concept or an analogous term. The case of being is intrinsic.

A thing is analogous by proportionality when we can say that it is or means what was suggested to some extent.[4] Analogy of proportionality can be proper or improper, depending on whether the analogy is metaphorical or real.

From those dictionary definitions of economy, we can immediately distinguish two sets of meanings: first, a domain of human reality and activities—associated with the management of resources, income, commodities, and wealth to satisfy human needs and desires—and second, a way of performing these activities—efficiently. The first set is broader than the latter, which is a specification of the former that may or may not actually apply: there are both efficient and inefficient economic activities.

The first and broadest meaning refers to the human domain of satisfying human needs. We can affirm that actions aimed at the acquisition, possession, and use of goods for the satisfaction of human needs are economic. For French anthropologist Maurice Godelier (1966: 23), the "economic" includes all decisions and actions aimed at the satisfaction of human, material or spiritual needs, measured in material terms. For goods to be truly economic, they must be exchangeable (and, consequently, susceptible to being assigned an economic value), a characteristic that points to the social dimension of the economy. At the same time,

[4] On analogy, see for example Ralph M. McInerny (1961).

economic decisions and actions are free and often related to future or uncertain events. Given that needs are not univocally determined, people are free to choose the way to fulfill them: we need to eat, but we can choose the menu. In addition, if needs were fixed, economics would be "resolved" by technique alone. Therefore, setting economic ends as if they were simply given data is inadequate. Exercising a choice among ends is part of economics. The indeterminate nature of needs or ends is a condition of "the economic." Freedom, singularity, and the fact that many economic events happen in the future and are often difficult to predict all add uncertainty as another characteristic of the economic. This topic has been explored at length by economists like Frank Knight (in his 1921 work *Risk, Uncertainty and Profit*) and John Maynard Keynes (in his 1936 *General Theory*, among other writings).

By considering all these characteristics, I intend to define the economic in this broad sense as the field of human interaction associated with the acquisition, possession, use, consumption, production, and distribution of goods for the satisfaction of human needs and desires. This broad, literal meaning of "the economic" determines a portion of human reality, a specific human field or domain.

The second meaning is more specific or strict. It introduces a particular way of undertaking economic actions: the "efficient," "best possible," "optimal," or "maximizing" manner—or "the economic principle," as economists call it. Given that this more specific meaning of "the economic" has to do with the best way of using means—that is, an efficient allocation of resources to satisfy human wants—it may be safe to say that it is somehow more "perfectly economic." However, it is not necessary for all "the economic" to respond to this specific or strict sense. The relationship between these two notions represents a case of proportional analogy. The broad sense deals with "the economic," and the strict sense does so "economically." An economic action as, for example, buying or selling a product, is economic regardless of whether it is or is not economically performed. However, it is *doubly* "economic" if it is performed efficiently. Robert Scoon has distinguished the adjectives "economic" and "economical" in the same way, and he believes that they should not be confused; he maintains that the "economic" should not be reduced to the "economical" (1943: 311). Still, it seems that it is ordinarily better to perform economic actions economically, except when other motivations indicate the convenience of doing them non-economically.

In my earlier book (2020: 23–26), I have also shown that Carl Menger, John Neville Keynes, and Herbert Simon have made a distinction between these two meanings of the economy. Menger separates a "technical economic" (1923: 73) from an "economizing" economy (1923: 74, 76). The former aims at providing the goods that we need, and the latter, when means are insufficient, seeks to do so by "economizing" in the best possible way. We cannot identify, Menger states, the concept of "economy" (Wirtschaft) with the notion of "economical" (Wirtschaftlichkeit, 1923: 61). Thus, he affirms, it is not paradoxical to speak of an "economic economy" and of a "non-economic economy" (1923: 61). As I noted in my 2003 article, for Menger, in real economic acts, we do not have only economically pure reasons but also "error, ignorance, and external compulsion" ([1883] 1985: 64). Additionally, Menger identifies freedom of the human will as one of the elements that make a difference between the economic theory and the real world (cf. [1883] 1985: 214). He states that "by *economy* we understand the precautionary activity of humans directed toward covering their material needs; by national economy, the social form of this activity" ([1883] 1985: 63, italics in Menger´s text). This definition matches the broad notion of the economy.

Concerning Neville Keynes, he defines "economic activity" as a human activity directed toward the production and appropriation of exchangeable means to satisfy human needs ([1890] 1955: 99–100), while an economic action "attains its end with the least possible expenditure of money, time and effort" ([1890] 1955: 1). Finally, Simon considers economic behavior as a satisfying rather than a maximizing behavior, with the latter as a subset of the former. He defines economics as concerned "with a particular subset of man's behaviour"—the first meaning—and he critically refers to today's vision of the economy as reduced to the second meaning (1978: 1).

This distinction between a broad and a narrow or focused concept of the economy applies to the microeconomic and macroeconomic realms. Contemporary economics deals with the narrow concept of economy at both the micro and macro levels. Yet, given that economic actions do not necessarily accomplish maximization, economics implicitly becomes normative rather than explicative or descriptive.

It is time now to consider the notion of the economy held by the metaphysicians mentioned in the previous chapter—Aristotle, Cartwright, Lawson, and Mäki—and their metaphysical characterizations of the economy.

Aristotle on the Economy[5]

Aristotle stated seminal concepts on "the economic." Evidently, the economy of his time was not like today's, and economics as a subject had yet to emerge; he only devoted a few pages to these issues. However, in those pages he left some ideas that may help clarify basic notions of the philosophy of "the economic."

First, I want to clarify the scope of the economy for Aristotle. Most historians of economic thought correctly translate *oikonomiké* as household management and thus consider that Aristotle's contribution to economic analysis was unimportant. Nonetheless, Aristotle held that *oikonomiké* ("the economic") refers not only to the household but also to the *polis* (cf. *Politics* I, 8, 1256b 12–4; I, 10, 1258a 19–21; I, 11, 1259a 33–6). *Oikonomiké* is the Greek adjective commonly used by him to refer to all that is related to the use of wealth to achieve the "good life," which is a life of virtues that makes people "flourish" by achieving *eudaimonia*, which for Aristotle is more perfect than happiness in its ordinary hedonic sense. This notion corresponds to the broad concept of the economy. For Aristotle, the narrow concept, as I will explain, refers to a "justly censured" form of *chrematistiké*, the art of providing the means that *oikonomiké* uses.

The adjective *oikonomiké*—economic or "the economic"—calls for a noun: the economic what? According to Aristotle, the answer is multiple: "economic" applies to several things and not univocally. In this case, as will be shown, we are facing an analogy of attribution. This conclusion clearly arises when comparing the focal meaning of *oikonomiké* with its other meanings; these different meanings pertain to different entities to which the adjective applies. These entities are listed below.

1. A Human Action

Let us begin with the "focal" meaning. The "focal" meaning of "the economic" for Aristotle is likely to be found in his definition of the economic. We shall confirm this hypothesis when we compare it with other entities that he also calls economic. Aristotle defines *oikonomiké* by relating it to *chrematistiké*. *Oikonomiké* is the use (*chresasthai*) of wealth, while *chrematistiké* is the provision, production, or acquisition of wealth.

[5] Part of this section is extracted from my article 2006.

"Using" is a human action. Thus, economic is for Aristotle a human action: using wealth. However, *oikonomiké* does not suggest unlimited wealth, but the wealth necessary to live at all (*zên*) and to live well (*eû zên*)—the good life (cf. *Politics* I, 4, 1253b 24–5). Aristotle also considers chrematistics as a human action: a technique that ought to be subordinated to *oikonomiké*, dealing, as noted, with the acquisition of things used by *oikonomiké*. In addition, he distinguishes two kinds of chrematistics: one subordinated to *oikonomiké*, limited and natural, and another unnatural one that in fact is not subordinated to *oikonomiké* and looks for money endlessly. Concerning the former, he says:

> It follows that one form of acquisition is naturally a part of the art of household management. It is a form of acquisition which the manager of a household must either find ready to hand, or himself provide and arrange, because it ensures a supply of objects, necessary for life and useful to the association of the polis or the household. (*Politics* I, 8, 1256b 27–30)

Concerning the latter, he states, "this second form [leads] to the opinion that there is no limit to wealth and property" (*Politics* I, 9, 1257a 1). He describes it, as mentioned, as "justly censured" (*Politics* I, 10, 1258b 1).

Thus, for Aristotle, *oikonomiké* is the action of using the things that are necessary for life (to *live at all*) and for the good life (to *live well*). When Aristotle speaks about life at all, he is referring to what is achieved at home (*oikos*). When he talks about the good life, he means what is attainable in the civil community (*polis*) and to its goal (*telos*). According to him, the latter concept of life has a precise moral meaning; it is, as mentioned, a life of virtue by which humans flourish. Summing up, this first and primary meaning of *oikonomiké* is the action of using the things necessary for life and for the good life.

Finally, we need to clarify what kind of action the economic is. In *Metaphysics,* Aristotle distinguishes between two kinds of human actions. First, *immanent* actions are actions whose end is the action itself, such as seeing, thinking, or living. The results of immanent actions remain in the agent. Second, he notes that, in *transitive* actions, the "result is something apart from the exercise, (and thus) the actuality is in the thing that is being made" (*Metaphysics* 1050a 30–1). Transitive actions are actions whose results transcend the agent and are something different from the agent, as, for example, a good produced. Aristotle uses the term *prâxis* to refer to immanent actions and *poíesis* to refer to transitive actions (*Nicomachean Ethics* VI, 4, 1140a 1). All actions are both immanent and transitive, except fully immanent actions (to think, to love). Let me provide an example: when somebody works, there are two results: an "objective" result, such as a product or service (transitive), and a "subjective" result, such as that individual's increased ability or self-fulfillment, as well as the morality of the act (immanent). For Aristotle, this latter, immanent aspect is the most relevant one; it is the one sought for its own sake, not for any other reason. Aristotle says, "we call that which is in itself worthy of pursuit more complete than that which is worthy of pursuit for the sake of something else" (*Nicomachean Ethics* I, 7, 1097a 30–1). That is, Aristotle attributes more relevance to the inner or immanent aspect of an action—that which is in itself worthy of pursuit—because it is the aspect whose *telos* is the agent's very fulfillment or perfection. For him, the external aspect of an action is simply instrumental.

Oikonomiké is an action of using, in Greek, *chresasthai*. What kind of action, immanent, or transitive, is *chresasthai*? "*Using*" is a transitive action insofar as the thing used is consumed or wasted when used. Still, the complete action of *oikonomiké* is to use what is necessary to satisfy the agent's requirements to live well: this is a predominantly immanent consideration of use, for it is using for the sake of personal perfection.[6] This conclusion confirms the essentially ethical character of Aristotle's *oikonomiké*.

2. A Capacity

Aristotle states: "(…) and we see even the most highly esteemed of capacities (*dynámeon*) to fall under this [Politics], for example, strategy, economics (*oikonomikén*), rhetoric" (*Nicomachean Ethics*, I, 2, 1094b 1–2). That is, he also considers *oikonomiké* as a capacity, an ability, or power, in this case, to perform economic actions. According to Aristotle, abilities or *dynameis* are powers to do. His definition in *Metaphysics* is similar to the definition of nature (*physis*): "a source of movement or

[6] *Chresasthai* is the nominalisation of the Greek verb *chráo* in its middle voice infinitive aorist form. The middle voice has a reflexive use that is coherent with this possible predominant sense of *prâxis* of *chresasthai*. The French and Spanish translations show this characteristic: "se server" (Fr.)/"procurarse de," "servirse de" (Sp.). *Chresoméne*, another form used by Aristotle to refer to the action of *oikonomiké* is another form of *chráo*, a future middle participle that indicates finality.

change, which is in another thing that the thing moved or in the same thing *qua* other" (V, 12, 1019a 15–6). *Dynamis* is an "urge of nature to grow to maturity, to realize form [essence], and to perform the due function" (Guthrie 1967: 140).[7] An ability, for Aristotle, may also be a habit or disposition (*Categories* VIII).

Oikonomiké as ability is a derived sense of *oikonomiké* because the ability to use exists for the sake of the action of using. It is the ability to look for the goods that we need for the good life. Given that abilities are defined by their ends or functions (*De Anima* II, 4, 415a 16–21), these ends are ontologically prior to the very abilities and correspond to the primary meaning in a case of an analogical term such as *oikonomiké*. "The excellence of a thing is relative to its proper function," says Aristotle (*Nicomachean Ethics* VI, 2, 1139a 17). There are three levels: the ends (contributing to the good life), abilities, and actions.

3. A Habit

It seems reasonable that if *oikonomiké* is both an action and the ability to perform it, it also creates a habit that facilitates this action. For Aristotle, habits rely on natural dispositions and are propelled and reinforced by education and law. The very repetition of the action also consolidates the habit, thus building a kind of virtuous circle: actions-habits-actions. It also makes sense to find that *oikonomiké* is a habit that facilitates the immanent aspect of action—not a *téchne*—i.e., a habit of production. Indeed, Aristotle speaks about household management as a kind of prudence, which in the Aristotelian conception mainly reinforces human beings' immanent ability (*Nicomachean Ethics* VI, 8; cf. also *Eudemian Ethics* I, 8, 1218b 13).

Oikonomiké as a kind of habit is another derived sense of *oikonomiké*. The same argument stated above—*oikonomiké* as ability being a derived meaning—applies in this case: the primary meaning, to which this derived meaning is oriented, is the proper object of the habit—that is, the corresponding action. *Oikonomiké* as a kind of habit helps the performance of *oikonomiké* as the action of using the necessary things to live well. Also, *chrematistiké* is a technique, which is a habit of production for Aristotle

[7] According to the Greek-English Lexicon of Lydell–Scott (Oxford, Clarendon Press, 1900), *dynamis* is a power, might, and strength; an ability to do something, a power, faculty, and capacity.

(cf. *Politics* I, 9 and 10, e.g., 1257b 7). The action of *chrematistiké* is transitive.

4. A Science

For Aristotle, *oikonomiké* is also a science (cf. *Nicomachean Ethics* I, 1 and 2), a practical science. Aristotle distinguishes between theoretical, practical, and *poietical* (or technical) sciences. This distinction corresponds to their different subjects (*Metaphysics* VI, 1, 1025b 20–1 and cf. 1025b 19ff. and XI, 7, 1063b 36–1064a):

1. The strictest notion of science for Aristotle, a theoretical science deals with things that can only be contemplated. For Aristotle, theoretical sciences include Metaphysics, Physics, and Mathematics.
2. A practical science deals with subjects that originated in human decisions or choices. They have a practical aim (*Metaphysics* II, 1, 993b 21–2; cf. also *Nicomachean Ethics* I, 2, 1095a 6 and II, 2, 1103b 27–8).
3. A technical science deals with artifacts and the rules for their production.

Politics is the Aristotelian practical science *par excellence.* Ethics and *oikonomiké* are also practical sciences for Aristotle.[8] Practical sciences are sciences by similarity: *kath' homoiótesin* (cf. *Nicomachean Ethics* VI, 3, 1139b 20). This is a middle ground between strict sciences (theoretical), and prudence and action. Consequently, this homonymous meaning of science is not the clearest and most central one. However, practical sciences feature a characteristic common to all kinds of sciences—i.e., to be a "state of capacity to demonstrate (*héxis apodeiktiké*)," (*Nicomachean Ethics* VI, 3, 1139b 32) with the limitations inherent to their subject matters, human choice, and human action (contingent, variable, free, singular). This last meaning of *oikonomiké* as practical science is also analogue—homonymous *pròs hén*—to "economic" human action. I will expand on *oikonomiké* as a practical science in Chapter 10.

[8] For *oikonomiké* as a practical science, cf. Newman (1951, I: 133), Miller (1995: 6–11), Natali (1980: 115ff.) and Berti (1992: 80). Instead, *chrematistiké* is a *poietical* science or technique.

Summing up, this property—*oikonomiké*: "the economic"—applies to different though related realities, such as economic action, economic capacity, economic habit, and economic science. They have something in common: their orientation toward the use of things necessary to live well, but they differ in other aspects—properly a homonymous *pròs hén* concept.

It is time now to apply Aristotle's metaphysical categories to these meanings of the economy (or economic). I have characterized *oikonomiké* grammatically as an adjective. Nouns may express both substances (things) and accidents (properties)—for example, "the earth" and "the beauty"—while adjectives nearly always denote accidents. In this case, the adjective *oikonomiké* is clearly not expressing something ontologically separable (i.e., a substance): its existence can only be explained or sustained by invoking something else in which it inheres.

The four kinds of entities that may be "economic" for Aristotle—an action, an ability, a habit, and a science—are accidents or properties, pertaining to the categories of "action" and "quality." They inhere in or "happen" to human beings (and somehow to firms, organizations, nations). The focal meaning of *oikonomiké* is the action of using. However, in order to use, people need to know what to use and how to use it, and they also need to have the ability to use; by using, people develop the corresponding habits. Let us analyze specifically what kinds of categories—ways of being—these meanings of the economic are.

(i) *Action* belongs to the category of action: *Categories* IX. Human action—*praxis*—is the most perfect "sub-lunar" way of being of actuality or *energeía* (cf. *Metaphysics* IX, 6). Humans try to achieve perfection through action. This is a reason why *oikonomiké* is a typically human entity. Previous activities needed to act—i.e., deliberation and choice —are qualities of the mind and the will. The use of wealth is a kind of human action. As noted before, it has both an immanent and a transitive character. Human actions are voluntary and intentional. They do not just only happen to humans, as if they were something alien to them: they presuppose previous activities in the same person. Some of these activities are intellectual—knowledge, belief—and other, volitional—will, choice, and decision. Aristotle considers deliberation of mind (*bouleúesthai*) and choice of will (*proaíresis*) as the previously required acts preceding action. Science, ability, and habit facilitate these previous steps.

As explained, economic action is for Aristotle the action of using the things necessary to live and to live well (in a moral sense). It is subjective because each person judges what is necessary for herself. The Greek term used in this case by Aristotle to mean necessity is *chreia* (cf. *Nicomachean Ethics*, V, 5). There is another Greek term for necessity, *anagke*, also used by Aristotle in other contexts. *Anagke* is a strict necessity (as, for example, it is necessary that an effect has one or more causes), but *chreia* is relative necessity: to survive, it is necessary to eat, but one may eat one food or another, depending on the time of day, and so on. Referring to *oikonomiké*, *chreia* means that the way of using the things required is not determined a priori, but it is up to everyone's will. This characteristic of "the economic" reinforces its accidental character—that is, "the economic" does not have a concrete determined content (i.e., it is contingent).

(ii) *Capacity* (*dýnamis*), to have a power ("a source of movement or change": *Metaphysics* V, 12, 1019a 15) is a quality. Capacities, for Aristotle, are natural (*physikes*) (*Categories* VIII 9a 14ff.). A capacity is an ability, potentiality, power, or talent possessed, in this case, by a human being. Human nature is equipped with some capacities that require development. Capacities may be dormant or active. Other capacities are not innate but acquired. *Oikonomiké* is one of these, related with rationality, which is innate but has to be developed. This characteristic of capacities reinforces their accidental character.

(iii) Habit (*héxis*) is also a quality, a "having" (*Metaphysics*, V, 20). Habits are more lasting and stable qualities than dispositions. Virtue (*areté*) is a quality also belonging to the sub-type of habit (*Categories* VIII 8b 34–5). Virtues are built on a natural disposition through repetition of actions. A habit is an acquired behavior pattern regularly followed until it has become almost involuntary or a dominant or regular disposition or tendency. The personality is shaped by acquiring habits: they constitute a person's "second nature." Habits are determined by actions, but actions are free. Thus, they may differ from one person to another. Hence, habits are accidents, and they are also contingent. In the social realm, habits are embodied in institutions.

(iv) Knowledge and science are qualities (*Categories* VIII, 8b 29–33), concretely a kind of habit. As explained, *oikonomiké* is a practical science, and this kind of science is not an exact science: the truth of the practical is not fixed.

As anticipated, these four meanings of *oikonomiké* correspond to accidents of the human being, which is substantial. I will extract conclusions from this in due time. Let us move on to the other metaphysicians now.

Cartwright on the Economy

Cartwright indirectly reveals her conception of the economy when analyzing the difficulties to work on economic theories: causes' instability, the relevant influence of structural factors, and the general characteristics of the social world that apply to the economy—complexity, reflexivity, lack of control, and singularity. She asserts, "Most of what happens in the economy is a consequence of the interaction of a large number of factors" (Cartwright 2001: 279). Circumstances leading to particular combinations affect the stability of causes. Without stable causes, we do not have any capacities, which for her, are necessary to formulate theories. In contrast, she notes that J. M. Keynes conceives of a world of causes but not of capacities (1989: 157). That is, the problem is not the absence of causes but their instability. This is why she asserts that her claim is "not that phenomena of economic life are governed by capacities, but rather that the method for econometrics presuppose this" (1989: 158).

For Cartwright, the results of economic events depend on structural circumstances. She states:

> The natural thought about the difference between the most fundamental capacities studied in physics and the capacities studied in economics is that the economic capacities are derived whereas those of fundamental physics are basic. Economic features have the capacities they do because of some underlying social, institutional, legal and psychological arrangements that give rise to them. So the strengths of economic capacities can be changed, unlike many in physics, because the underlying structures from which they derive can be altered. (2007a: 54)

She then suggests that we should try to understand how these structures affect outcomes (2007a: 79, 2009: 57).

In conclusion, given that economic causes are highly dependent on structural circumstances and that we do not have many principles in economics, we need to find out how these circumstances influence the outcomes. These outcomes will consequently be "special-case conclusions, not general claims about the results of the capacity" (2007a: 75,

2009: 50). I will revisit this conclusion in Chapter 10, which deals with the nature and methodology of economics. What matters for this chapter is that Cartwright remarks the dependence of economic affairs on a series of factors that are not uniform. Consequently, for Cartwright, the economic theory is highly dependent on these factors that are contingent, and their correlation with reality implies the need to circumscribe it to local situations.

Lawson on the Economy

In the Introduction to this chapter, I have lamented that Lawson's books included no definition of "the economy." However, there is a chapter about Economics in *Reorienting Economics* (2003: Chapter 6). It considers several traditional definitions of the subject of economics, claiming that he will try to reconcile them rather than declaring them incompatible.

This "reconciliation" must be considered in the context of this author's ideas on ontology—more specifically on social ontology. I have explained that Lawson maintains that there is a level of reality that runs deeper than observable phenomena, consisting of underlying causes, structures, powers, mechanisms, and tendencies (2003: 143–144). Concerning the social realm, he sustains that there are "structures whose existence depends upon human intentional agency, underlying and influencing manifest phenomena such as human actions as well as states of affairs" (2003: 147). These structures are rules, relationships, and positions. The link between human agency and those structures is bi-directional. Lawson states, "All social structure is produced, reproduced and transformed through human practice, necessitating that structure and practice be conceived somewhat holistically as two essential, if distinct, aspects of the same process" (2003: 149).

This is why social reality is necessarily historically and geographically based. Social phenomena, in addition, are emergent and possess emergent properties. They are not reducible to the lower level but are conditioned by and dependent on it.

Concerning the economy, Lawson thinks that it is an inseparable part of the social realm. He states:

> I cannot think of a single sphere of human activity—from lending support to a football team, to listening to music, or even to making love—that does

> not (or could not) have an economic aspect (given the interpretation of economics set out above.). These and all other activities take place in space and time, both of which can have alternative uses. All human activities require material conditions and are, for example, influenced by property relations, the length and structure of the working "day" or "week," opportunities forgone, and so on. But at the same time very few activities, if any, have merely an economic aspect, including those of labouring on the shop floor or in an office or doing shopping. Activities such as the latter can involve wearing the colours of, and so signalling support for, a particular football team, or, say, being on the lookout for an interesting companion or just being part of a crowd/community, amongst many other things. (2013: 162)

Within this conception, Lawson encompasses the subject matter involved in the definitions of economics crafted by Mill, Marshall, and Robbins. All human activity is economic in some sense, as Marshall implicitly notes in his definition:

> Political Economy or Economics is a study of mankind in the ordinary business of life; it examines that part of individual and social action which is most closely connected with the attainment and with the use of the material requisites of well-being. ([1890–1920] 1962: 1)

Mill's definition concerns the causes of wealth and the conditions of human activity associated with production and well-being. Finally, Lawson considers Robbins' definition—the optimal relationship between ends and scarce means which have alternative uses (1935: 16)—as a practical aspect (not necessarily essential) of economic behavior or as a normative ideal.

Finally, Lawson argues that the economic dimension is present in all social phenomena. I do not share this idea, which resembles the views of the economic imperialism supporters criticized in Chapter 1. I conceive the economy as a sector of the human social realm, with the economic approach—maximizing—as limited to this sector.

Mäki on the Economy

As I pointed out, Mäki has not defined the economy, because his interest largely focuses on the ontology of economics. However, we can find some ideas about the characteristics of the economy in his works. First, I have already mentioned that, for him, economics deals with commonsensibles, which are not necessarily material: "costs and benefits, tastes and choices, firms and households, money and the market" (2002: 95). In his early works, he spoke about "folk economic views," comparing them with Aristotle's *oikonomiké* (e.g., 1996: 436–437): economic phenomena are intentionally brought about by people in charge, like governments, firms, and—I add—ordinary people.

Second, these commonsensibles are mind-dependent. Mäki thinks that the empirical and social criteria for the election of economic theories are limited. He also believes that the ontological criterion consisting in "the way the world works" is more relevant, and that this is a matter that hinges on causal processes (2001a: 371; see also 2009: 39). He also speaks of causal powers and mechanisms (e.g., 2008).

Like Lawson, Mäki views the subject matter of economics as combining Marshall's and Robbins' definitions. He asserts (2001b: 4):

> We may think of the economic realm as being distinguished from other realms (the physical realm, the biotic realm, the realm of aesthetics) by virtue of being composed of certain types of entities and properties—such as "that part of individual and social action which is most connected with the attainment and with the use of the material requisites of wellbeing" (Marshall 1920, 1) or "human behaviour as a relationship between ends and scarce means which have alternative uses". (Robbins 1935: 16)

He also holds that the possibility of the existence of the economic realm somehow depends on the existence of the physical, the biotic, and the psychics realms, and that the "functioning of the economic realm is shaped by the realms of morality and politics" (2001b: 4). He distinguishes ontological categories that range from specifically economic, more general in applicability, and finally, universal. It is interesting to quote his list, because it is, in fact, his ontological proposal—much like Varzi's (the inventory of economic entities already quoted from 2011 in Chapter 2)—and also includes the most general metaphysical categories (2001b: 8):

> We get a continuum of sets of categories in an ascendant order of generality: money, market, firm, price; preference, belief, rationality, choice, rule, welfare; evolution, aggregate, equilibrium; quality and quantity, probability, essence, causal power, mechanism, existence, objectivity.

Summing up, though he does not define the economy, Mäki sheds a light on some of its more interesting characteristics.

Conclusion: The Nature of the Economy

After exploring the meaning of the economy in several dictionaries, I concluded—in agreement with Menger, Neville Keynes, and Simon—that we can establish two complementary notions of it. First, the term economy refers to a broad concept that defines an area of human interaction associated with the exchange, production, and distribution of goods to satisfy human necessities and desires. Second, its restricted notion defines an efficient way of doing those activities. In this case, we have an analogy of proportion.

Instead, the analysis of Aristotle's notion of the economy concludes that it is a homonymous or analogue-by-attribution concept, whose focal meaning is economic action: the use of what is needed to live and to achieve the good life. The other meanings of "the economic" are a capacity, a habit, and a science. In all its four meanings, we refer to accidents (properties), entities which have an ontological existence inhering in others—which are substances (things) or function as if they were substances. It is not contradictory to consider the two notions of the economy mentioned in the previous paragraph as proportionally analogue, and Aristotle's notion as analogue by attribution, because Aristotle's *oikonomiké* did not include the strict notion of economy.

For Cartwright, we must ascertain the capacities or natures—stable causes—of the things studied to do science. She thinks that the problem with the social realm and specifically with the economy is the "paucity" of these stable causes. Economic causes are highly dependent on structural circumstances—social, cultural, institutional, legal, and psychological factors that are contingent—and their correspondence with reality requires circumscribing them to local situations.

Lawson also remarks the need to delve deeper into reality to look for the structures, causes, capacities, and tendencies that lie beneath observable phenomena. Therefore, for him, social reality is necessarily historical and geographical. Social phenomena, in addition, are emergent and possess emergent properties. He thinks that the economic dimension is present in all social phenomena. As noted earlier, I do not agree because I view the economy not as a dimension but as a part of the human realm.

Finally, according to Mäki, economics deals with mind-dependent commonsense entities. This mind-dependence seems to separate him from Lawson, who appears to hold a materialist conception. Mäki remarks that we must look for the causes of economic entities, upholding that the existence of the economic realm depends on the existence of the physical, the biotic, and the psychics realms. He also believes that its functioning is shaped by the realms of morality and politics.

Combining the ideas presented above—in which I find a lot of intersections—I extract the following conclusions:

1. The fact that for Aristotle all his related categories that characterize the nature of the economy—action, capacity, habit, and science—are properties makes it dependent upon the traits of the subject in which economic entities inhere—that is, human beings and the society.
2. Cartwright does not define the economy or economic entities, but she stresses that they are dependent on unstable causes, the interaction of many factors, and underlying structural arrangements that are not always the same.
3. Despite my disagreement with Lawson's definition of the subject matter of economics (2003: 162–164), it should be noted that he intimately connects it with the whole human life, sharing its underlying structures that are historically and geographically variable.
4. Mäki combines Marshall's and Robbins' definitions of the economy (roughly corresponding to the broad and narrow conceptions of it) and makes them dependent on causal powers and mechanisms. He also thinks that the economic realm depends on the physical, biotic, psychical, moral, and political realms.
5. In conclusion, the economic realm—the part of human reality dealing with the goods that satisfy human needs and desires—consists of accidental entities or properties that cannot be separated from underlying variable structures, capacities, and institutions, or from the agents that have diverse motivations influencing economic actions. This contingent realm calls for local analyses.
6. These conclusions have an impact on the epistemological status and methodology of economic science. This is the topic of Chapter 10.

The Ontology of the Economy

The ontological question is which the economic entities are. As mentioned in the previous chapter, Mäki states that the ontology of the economy should answer these questions: "What is the economy made of? What are its constituents and how do they hang together?" (Mäki 2001a: 3).

Mäki argues that "Economic theory talks about the common sense furniture of the human world: costs and benefits, tastes and choices, firms and households, money and market" (2002: 95). Economics then elaborates on them (not always appropriately). For Mäki, "commonsensibles" are not necessarily observable but "experienced" by means of "introspection, inference, interpretation, culturally established meanings" (1999: 248). The "commonsensible" condition of economic entities seems to suggest that it will be easy to ascertain their meaning. However, this does not preclude the possibility that there are also non-commonsensible economic entities.

I will try to organize this constellation of constituents of the economic realm.

- First, we have the substance or thing around which all the economy is built: human beings, who have an economic dimension—i.e., economic agents.
- Economic agents perform economic actions: buying, selling, saving, investing, borrowing, and building a capital or wealth, working, producing, and distributing—all actions that are accidents or properties of economic agents in a broad sense.
- All these actions ultimately are directed to the production, acquisition, and consumption or use of economic goods, which are substantial things or services.
- Because they are economic, these goods have a value—one of their properties—expressed by a price in a monetary economy.
- The division of labor—a human action of economic agents—allows or is propelled by the exchange of goods in the market.
- Some institutions appear to perform or regulate those activities: firms, banks, trade unions, government economic institutions, and regulations…
- From the interaction of all these players emerges a whole economic system, a macroeconomic entity.

As anticipated in its Introduction, this book will not deal with the nature of all these economy constituents. This chapter has dealt with economic actions in general, and the rest of the book will focus on analyzing economic agents' identity (Chapter 5), workers' motivations (Chapter 6), the nature of macroeconomic entities and their relationship with microeconomic entities (Chapter 8), and, finally, money (Chapter 9).

References

Aristotle. (1995). *The Complete Works of Aristotle. The Revised Oxford Translation*, J. Barnes (ed.) Princeton: Princeton University Press, 6th printing with corrections.

Berti, E. (1992). *Le vie della ragione*. Bologna: Il Mulino.

Cartwright, N. C. (1989). *Nature's Capacities and Their Measurement*. Oxford: Oxford University Press.

Cartwright, N. C. (2001). "Ceteris aribus Laws and Socio-economic Machines". In: U. Mäki (ed.) *The Economic World View: Studies in the Ontology of Economics*. Cambridge: Cambridge University Press, pp. 275–292.

Cartwright, N. C. (2002). "The Limits of Causal Order, from Economics to Physics". In: U. Mäki (ed.) *Fact and Fiction in Economics: Models, Realism and Social Construction*. Cambridge: Cambridge University Press, pp. 137–151.

Cartwright, N. C. (2007a). *Causal powers: What are they? Why do we need them? What can and cannot be done with them?* Contingency and dissent in science series. Centre for Philosophy of Natural and Social Science, LSE, London. http://www.lse.ac.uk/collections/CPNSS/projects/ContingencyDissentInScience/DP/CausalPowersMonographCartwrightPrint%20Numbers%20Corrected.pdf. Accessed 20 June 2018.

Cartwright, N. C. (2007b). *Hunting Causes and Using Them: Approaches in Philosophy and Economics*. Cambridge: Cambridge University Press.

Cartwright, N. C. (2009). "If No Capacities Then No Credible Worlds: But Can Models Reveal Capacities?", *Erkenntnis* 70: 45–58.

Crespo, R. F. (2003). "Three Arguments Against Menger's Suggested Aristotelianism", *Journal des Economistes et des Etudes Humaines* 13/1: 63–84.

Crespo, R. F. (2006). "The Ontology of 'The Economic': An Aristotelian analysis", *Cambridge Journal of Economics* 30/5: 767–781.

Crespo, R. F. (2011). "Two Conceptions of Economics", *Journal of Applied Economics*, XIV/2: 181–197.

Crespo, R. F. (2013). *Philosophy of the Economy: An Aristotelian Approach*. Dordrecht: Springer.

Crespo, R. F. (2020). *The Nature and Method of Economic Sciences: Evidence, Causality, and Ends*. London: Routledge.

Godelier, M. (1966). *Rationalité et irrationalité en économie*. Francois Maspero, Paris.

Guthrie, W. K. C. (1967). *The Greek Philosophers*. London: Methuen.

Keynes, J. N. ([1890] 1955). *The Scope and Method of Political Economy*. Fourth Edition. New York: Kelley and Millman.
Lawson, T. (2003). *Reorienting Economics*. London: Routledge.
Mäki, U. (1996). "Scientific Realism and Some Peculiarities of Economics". In: R. S. Cohen, R. Hilpinen, and Q. Renzong (eds.) *Realism and Anti-Realism in the Philosophy of Science*. Kluwer, pp. 427–447.
Mäki, U. (1999). "Representation Repressed: Two Types of Semantic Scepticism in Economics". In: R. Rossini Favretti, G. Sandri and R. Scazzieri (eds.) *Incommensurability and Translation*. Cheltenham: Edward Elgar, pp. 307–321.
Mäki, U. (2001a). "Economic Ontology: What? Why? How?". In: U. Mäki (ed.) *The Economic World View: Studies in the Ontology of Economics*. Cambridge: Cambridge University Press, pp. 3–14.
Mäki, U. (2001b). "The Way The World Works (WWW): Towards an Ontology of Theory Choice". In: U. Mäki (ed.) *The Economic World View: Studies in the Ontology of Economics*. Cambridge: Cambridge University Press, pp. 369–389.
Mäki, U. (2002). "Some Nonreasons for Nonrealism About Economics". In: U. Mäki (ed.) *Fact and Fiction in Economics*. Cambridge: Cambridge University Press, pp. 90–104.
Mäki, U. (2008). "Philosophy of Economics". In: S. Psillos and M. Curd (eds.) *The Routledge Companion of the Philosophy of Science*. London: Routledge, pp. 543–554.
Mäki, U. (2009). "Realistic Realism About Unrealistic Models", In: H. Kincaid and D. Ross (eds.) *The Oxford Handbook of the Philosophy of Economics*. Oxford: Oxford University Press, pp. 68–98.
Marshall, A. ([1890–1920] 1962), *Principles of Economics*, (8th edition, 1920), London: MacMillan.
McInerny, R. M. (1961). *The Logic of Analogy*. The Hague: Martinus Nijhoff.
Menger C. ([1883] 1985). *Investigations Into the Method of the Social Sciences with special reference to economics*, edited by Louis Schneider and translated by Francis Nock. New York: New York University Press.
Menger, C. (1923). *Grundsätze der Volkswirtschaftslehre*. Mit einem Geleitwort von Richard Schüller, aus dem Nachlass herausgegeben von Karl Menger. Wien: Hölder-PichlerTempsky/Leipzig: G. Freytag.
Miller, F. D. Jr (1995). *Nature, Justice and Rights in Aristotle's Politics*. Oxford: Oxford University Press.
Natali, C. (1980). *Aristotele e l'origine della filosofia pratica*. In: C. Pacchiani (ed.) *Filosofia pratica e Scienza Politica*. Padova: Francisci, pp. 99–122.
Newman, W. L. (1951). *The Politics of Aristotle*. Oxford: Clarendon Press.
Robbins, L. C. (1935). *Essay on the Nature and Significance of Economic Science*. London: MacMillan, Second Edition.

Schabas, M. (2009). "Constructing 'The Economy'", *Philosophy of the Social Sciences* 39/1: 3–19.

Scoon, R. (1943). "Professor Robbins' Definition of Economics", *Journal of Political Economy* 51/4: 310–320.

Simon, H. (1978). "Rationality as Process and as a Product of Thought", *American Economic Review* 68/2: 1–16.

Soler Miralles, J. E. G. (1952). "Sobre Filosofía de la Economía", *Revista de la Facultad de Ciencias Económicas*, UNCuyo, 12/September–December: 129–39.

Varzi, A. C. (2011). "On Doing Ontology Without Metaphysics", *Philosophical Perspectives* 25: 407–423.

CHAPTER 5

The Identity of the Economic Agent

This chapter deals with the nature of the individual economic agent. John Davis, in the first chapter of his first book on the notion of individuals in economics, anticipates that neoclassical and mainstream economics lack an adequate grasp of them, treating them "as relatively autonomous or even atomistic beings" (2003b: 17). He traces this misconstruction back to the influence of philosophers like René Descartes and John Locke, and he undertakes an ontological analysis in order to look for a notion of individuals in economics that matches reality. In effect, human beings are not atomic units, but they are strongly shaped by their social environment and culture.

Russian émigré to the United States and Harvard University Professor Pitirim Sorokin (1889–1968) proposed, among other rich contributions to the field of sociology, a doctrine that has been called "integralism." He believed that there is an "integral indivisible sociocultural trinity": society, culture, and personality. One of Sorokin's greatest contributions is the idea that society cannot be understood independently of culture and personality. He studied "integralism" in his book *Social and Cultural Dynamics* (1937–1941), while *Society, Culture, and Personality* (1947) provides a synthesis of Sorokin's scientific inquiries.

According to Sorokin, the three systems—social, cultural, and personal—interact in every social action. The social system depends heavily on the specific culture of its place and time. Individuals follow

R. F. Crespo, *The Nature of the Economy*,
https://doi.org/10.1007/978-3-031-02453-5_5

social norms and roles, their actions abide by their cultural values, beliefs, myths, and representations. The sociological tradition represented by Max Weber, Sorokin, Robert K. Merton, and Talcott Parsons, among others, features this conception of the dynamic interaction between the three systems.

Parsons, for example, asserts (1951:6):

> A social system is only one of three aspects of the structuring of a completely concrete system of social action. The other two are the personality systems of the individual actors and the cultural system which is built into their action. Each of the three must be considered to be an independent focus of the organization of the elements of the action system in the sense that no one of them is theoretically reducible to terms of one or a combination of the other two. Each is indispensable to the other two in the sense that without personalities and culture there would be no social system and so on around the roster of logical possibilities.

Max Weber ([1922] 1978: 24–45) also considers these three factors. As already mentioned in Chapter 2, he distinguishes four types of motives guiding social actions: instrumentally rational, value-rational, affective, and traditional motivations. According to Weber, an action is instrumentally rational when it seeks the adequate distribution of means to obtain the actor's ends. It is value-rational when it is determined by a conscious belief in the intrinsic value of some form of behavior. Affective actions are guided by the actor's affections and feelings, while actions are traditional when they are determined by adopted habits. Weber believed that, though a particular type of action may prevail in some specific kind of rationality, almost all human actions stem from several of these types. Instrumental rationality is all about means with a number of given ends, and the other three motivations define ends and are closely associated with individuals' social links and culture. We need means and ends to act. For value-rational, affective, and traditionally guided actions, the very action is an end, regardless of its consequences. A commonly quoted example is voting: though an individual vote does not affect the overall result of an election, people vote because they feel this action as a worthwhile obligation in itself.

"Standard" economics only considers a form of the first type of motives for human actions proposed by Weber—maximizing instrumental rationality—or reduces all the other types to this one.[1] For example, in *Epistemological Problems of Economics* ([1933] 1960) and referring to Weber's classification of motives for social action, Ludwig von Mises argues that all these motives are comprised in the instrumental rationality motive and can be maximized. Mises is changing Weber's underlying concept of the maximizing principle, which is empirical, turning it into a "metaphysical" principle ([1933] 1960: 83–85), in the sense that it does not bring any relevant information: all actions maximize, and no action deletes this characteristic. This conception leads to either an incomplete or an unhelpful analysis. Personality and culture are put aside, but, as Elena Aleksandrova et al (2020: 4992) state in reference to Sorokin's integralism, this is "a necessary direction in socio-economic research of modern society."

In fact, as already noted, a lot of the "anomalies" in standard economic theory—rational choice theory and expected utility theory—that emerged in experiments conducted over the past 30 years have forced economics to consider imports from non-economic sciences to explain them. Thus, some new research programs, like behavioral economics, evolutionary economics, neuroeconomics, or the capability approach, which take elements from other sciences, have gained traction.

A member of Sorokin's trinity—personality—has started to garner attention with the studies on the identity of the economic agent. Cultural economics considers another member.[2] The identity of the economic agent is a factor that deserves careful analysis, and identity economics accounts for a significant new approach. Davis believes that, beneath the crisis of standard economics' concept of rationality, lies its poor notion of individual identity (Davis 2011: 3). Identity economics captures the idea that the personal identity of an individual—an economic agent—is important to explain individual economic behavior. Personal identity

[1] Ben Fine explains: "To put it bluntly, from the marginalist revolution onwards, the issue of identity has been discarded in deference to the ill-defined utility-maximising individual. On this basis, a standard technical apparatus has been constructed, with utility and production functions to the fore. Rather than considering the nature of identity as such, on its own merits as it were, it is re-introduced to fit within the limits of that apparatus whatever its intrinsic quality in general and for the notion of identity in particular" (2009: 185).

[2] For more on cultural economics, see Bruno Frey and André Briviba (2021).

greatly influences people's decisions, including their economic decisions. As Davis states, "Economics and social science [...] must make use of some conception of the individual to explain economic life" (Davis 2003a: 22). The notion of identity is relevant to economics in that it provides a necessary theoretical or philosophical framework underlying the descriptions of individual economic agents. Why do we need such a framework? Because if describing an individual economic agent is somewhat problematic, then we can argue that such description fails to identify him/her. The agent might still be modeled mathematically, as in standard optimization analyses, but, unless the underlying description can be said reasonably to identify the agent, there is no reason to believe that such an analysis refers to any particular individual. Indeed, we must be able to justifiably say to "whom" a description applies if we are going to claim it is a realistic description.

The literature on identity and economics upholds that agents' descriptions fail to identify real people. Identifying the economic individual poses an issue. Standard economics endorses an atomistic conception of individuals, as I have mentioned that Davis argues (2003a, b).[3] However, an individual's multiple social commitments and culture shape his/her sense of identity. Therefore, the atomistic individual conception proves inadequate for economics. Moreover, behavioral economics has shown that individuals often make choices that are influenced by their contexts. Yet, while contexts change, individuals remain the same, and individual identity is forged from choices, experiences, culture, and circumstances. Thus, it is relevant to know a person's identity in order to know how he/she will act. In fact, Google, Facebook, and Amazon use algorithms to detect users' characteristics, identities, concerns, and tastes to offer them goods and services accordingly. Shoshana Zuboff (2019) views this practice as excessive, calling it "surveillance capitalism."

Referring to the need of ascertaining the nature of the agent's identity, Davis (2021: 4) asserts:

[3] Nitasha Kaul (2008: 123) characterizes this agent: "economic agents are usually by definition rational, self-interested (thus profit-motivated), utility-maximising (thus constantly calculating), amoral, disembodied and disembedded, abstract individuals. As a Hobbesian mushroom, this homo economicus (HE) is the quintessential Rational Economic Man (REM)."

> [i]t is important to bear in mind that what social science ultimately aims to do is give explanations of the social world which account for what causes what. That is, social science, like natural science, seeks to produce causal explanations of the world. Accordingly, ignoring one set of causal relationships associated with individual behavior not only runs the risk of producing poor science but also of distorting our understanding of people's behavior.

Consideration of an individual's social links provides George Akerlof and Elizabeth Kranton with the kickoff for identity economics. The *Quarterly Journal of Economics*, in its August 2001 issue, published their article entitled "Economics and Identity." They drew the definition of identity as "a person's sense of self" (2000: 715) from social psychology.[4] These authors asked how personal identity affects economic facts (2000: 716) and stated that "[i]dentity can account for many phenomena that current economics cannot explain" (2000: 715). They regard identity as "a new type of externality" (2000: 717).

The notion of identity is also present in Nobel laureate Amartya Sen's "capability approach." In addition to his philosophical training, Sen draws from authors like Martha Nussbaum (Nussbaum and Sen 1993) and Michael Sandel (Sen 1999) on identity. Economists-philosophers Alan Kirman and Miriam Teschl have also reflected on identity in economics. Davis, a leading figure in the field of philosophy of economics, has extensively worked on the notions of identity and economics.[5]

As the reader has probably already realized, the significance of identity and of the concept of identity for economics depend on the notions of personal identity and of the economy upheld. Actually, there are activities that would perfectly be—and in fact are—performed by machines. Technical demands are not very exacting, while the economy, instead, deals with human beings. Economic actions are human actions, which are more complex than technical actions. Consequently, in this chapter, I will first explore the conditions economics requires for a concept of identity, and, then, I will propose a notion of human personal identity. Finally,

[4] On the largely psychological roots of this notion, see Davis (2011: 72–75, 78).

[5] For an analysis of these theories, see Anton Mlinar and Crespo (2021: 197–200) and Crespo (2020: 52–56). Béatrice Boulu (2015) views personal identity as a socio-psychological construct. Simen Bagce and Ensar Yilmaz (2020) have developed a critical, updated analysis of the theories on the identity of the economic agent. See also the 2021, 1 issue of the *Forum for Social Economy*.

I will apply this notion of human identity to economics, looking for an adequate description of the identity of the economic agent.

A Notion of Economics and Requirements for a Corresponding Concept of Identity[6]

Kirman and Teschl (2004: 62) assert that "[t]he economic agent creates, builds, changes, and learns, is self-reflexive and evaluates her actions." I have discussed (2013: Chapter 2, and 4 of this book) the deep meaning of economic matters—or "the economy"—from a philosophical standpoint, characterizing economic reality as free, uncertain, and embedded in time. I have also noted its subjective character and its social entanglement, exploring three meanings of "the economic": (1) a metaphoric or improper meaning—human beings are "economic" insofar as they have needs that they can satisfy using material means; (2) a proper, broad meaning—all decisions and actions geared to the acquisition and use of the goods that satisfy human needs are economic: economic affairs, as they are commonly understood, regardless of their motivations; and (3) a proper, precise meaning—the maximizing character of the use of means in order to achieve ends with those decisions and actions is specifically economic. This last meaning matches the notion adopted by standard economics. Nonetheless, it lacks the richness implied in Kirman and Teschl's description of the actions performed by economic agents and in my characterization of economic reality. We need a theory of agent and identity fitting with descriptions of "the economic" according to its second meaning—a proper broad meaning—because the third meaning does not consider all the motivations that might underpin economic actions, and I have argued (2013: Chapter 2 and Chapter 4 of this book) that the "focal" meaning of "the economic" is economic action.

Speaking about political economy, John Stuart Mill implicitly considers the second and third meanings mentioned above. I will repeat three quotations from him already cited in Chapter 2. He first defines political economy as follows:

[6] I draw from my chapter of a book co-authored with Ivana Anton Mlinar (Anton Mlinar and Crespo 2021) for this and the previous sections.

> What is now commonly understood by the term "Political Economy" is not the science of speculative politics, but a branch of that science. It does not treat of the whole of man's nature as modified by the social state, nor of the whole conduct of man in society. It is concerned with him solely as a being who desires to possess wealth, and who is capable of judging the comparative efficacy of means for obtaining that end. ([1844] 2006: 321)

The last part of the last sentence anticipates the currently prevailing definition of economics—namely the third meaning of the economic mentioned above: the optimum allocation of scarce means in order to satisfy given ends. However, Mill is aware that this description of political economy involves a simplifying abstraction:

> All these operations, though many of them are really the result of a plurality of motives, are considered by Political Economy as flowing solely from the desire of wealth [...] Not that any political economist was ever so absurd as to suppose that mankind are really thus constituted. ([1844] 2006: 322)

Therefore, he finally emphasizes the need to consider additional motives for these "operations" in order to come to a correct explanation and prediction:

> So far as it is known, or may be presumed, that the conduct of mankind in the pursuit of wealth is under the collateral influence of any other of the properties of our nature than the desire of obtaining the greatest quantity of wealth with the least labor and self-denial, the conclusions of Political Economy will so far fail of being applicable to the explanation or prediction of real events, until they are modified by a correct allowance for the degree of influence exercised by the other causes. ([1844] 2006: 323, see also 326–327)

Mill serves as an example because, although he prompted the coining of the expression "homo economicus," he recognizes that it is an unreal abstract concept and that real agents' economic decisions can be greatly influenced not only by economic factors but also by a plethora of motivations.[7] Adam Smith shares this view, and, as Milonakis and

[7] On the origin and meaning of the expression "economic man," see John Persky (1995).

Fine (2009: 19) assert, Smith's theoretical edifice is "rich and multi-faceted, encompassing philosophical, psychological, social, historical and economic elements." The consideration of a plurality of motives for economic actions constitutes a central characteristic of the German Historical School of Economics. Schumpeter (1954: 177–178) remarks that a key feature of this school is that it recognizes that human actions, including economic actions, are not motivated by economic rewards only but are mostly guided by a "multiplicity of motives," stressing the need to concentrate more on individual correlations than on the general nature of events.

Max Weber (1949: 65–66) distinguished "specifically economic motives" (almost corresponding to Lionel Robbins' definition of economics) from "economically conditioned" events and "economically relevant" activities and situations that are not specifically economic. According to him, we find specifically economic actions only in "unusual cases" ([1922] 1978: 15). He considers that there are at least four types of social actions—already mentioned: "instrumentally rational" (specifically economic actions), "value-rational," "affectual," and "traditional"—and that it would be "very unusual" to find actions "oriented only in one or another of these ways" ([1922] 1978: 32).

That is, as conceived by classical economic thinkers, a real "economic action" involves not only the motivations considered by standard economics, but also "a plurality of motives," "other causes." These other motives or causes lie at the root of the characteristics of economic actions as described by Kirman and Teschl as well as myself: they are rational, psychological, sociological, cultural, historical, and ethical—i.e., consistent with the second meaning noted above.

Let us use an example to illustrate this point. Buying a new car is an economic action. When buying a car, you make calculations and economic comparisons among car models, taking into account their specific features. However, you may also feel loyalty to a brand, sympathy for the seller, or you may be used to buying cars from a single dealership; you listen to and take into account your wife's tastes and opinions; you might be influenced by the beauty of a specific car, and so on. In short, there are plenty of motivations involved in the transaction.

Thus, we need a concept of agency and identity that supports all the abilities and characteristics implied in the second meaning above. In fact, these traits do not call for an "economic" specificity in agents and their identity. Agents performing economic actions are entirely involved in these actions. A notion of "economic agent" only motivated by maximization makes sense only for the third meaning the economic; consequently, as the third meaning is an "idealization," we can only speak about an "economic agent" as a simplification. In the second meaning, there is no economic agent, but "simply" a human agent that must be considered in her completeness, performing an economic action. Similarly, we do not need a specific notion of identity for economic actions but "simply" a human identity—in all its richness. The car purchase example shows how all kinds of motivations influencing human agency are involved. Hence, it seems that an identity theory becomes necessary to grasp economic affairs, because the economic agency is essentially ***human*** agency. Indeed, the agent, the ***acting-who*** must be considered in all her wholeness.

Standard economists can be skeptical about the usefulness of considering identity in economics. This is understandable in the context of economics as currently conceived. However, viewing economics as a social science, with a methodology that leaves room for prudential reason to assess decisions stemming from an incommensurable plurality of motives, identity happens to be a central motive. This concept differs greatly from today's economics but draws closer to classical political economy. I believe that the "spirit" of this ancient original thought about economic life should be reestablished and that personal and social identity constitute a critical factor for consideration. However, there is room for both conceptions, as quoted from Mill. The next step is to define personal identity.

On Personal Identity[8]

Nowadays, views on personal identity are called "complex" and "simple," and they correlate to notions coined by Ancient or Modern thinkers. The book *Personal Identity. Complex or Simple?* edited by Georg Gasser and Matthias Stefan (2012) contains a nice synthesis and appraisal of these theories. These theories intend to explain the persistence of personality over time and obviously correspond to their metaphysical underpinnings.

[8] I draw from my 2017 book chapter and from the book chapter I co-authored with Ivana Anton Mlinar (Anton Mlinar and Crespo 2021).

Complex Theories and Their Criticism

Complex theories analyze personal identity in terms of simpler physical or psychological entities or events. Simple theories, instead, regard identity as basic and cannot be analyzed in terms of simpler components—identity is simple and analyzable. As explained by Gasser and Stefan (2012: 3):

> The complex view analyzes personal identity in terms of simpler relations. The fact that a person persists over time is nothing more than some other facts which are generally spelled out in either biological or psychological terms, or both. That is, the complex view takes talk about "what personal identity consists in" literally. It aims to provide necessary and sufficient conditions for personal identity, thereby reducing it to the holding of basic biological or psychological relations. Whenever these relations obtain, personal identity obtains.
>
> The simple view of personal identity, by contrast, denies that a person's identity through time consists in anything but itself. Biological and psychological continuity may be regarded as epistemic criteria for diachronic identity, but they are neither necessary nor sufficient conditions for personal identity. There are no non-circular, informative necessary and sufficient conditions for personal identity: personal identity consists in nothing other than itself.

The complex view includes two main approaches: a biological and a psychological approach. The former refers to the body–the so-called animalism—and to a relevant part of the body–the brain-based approach.

Animalism maintains that the continuity of human beings depends on the preservation of their biological organization. If, for example, the brain fails, and the human being continues to live, the latter will not be a person but a human organism.[9] In this approach, the identity pertains to the animal organism, not the person. With the brain-based approach, instead, the persistence of the brain is essential for the continuity of the human person's identity.

The psychological approach initiated by John Locke views the human person's persistence as dependent upon the connection of psychological states: we are unified bundles of mental states. Most contemporary philosophical discussions on personal identity (Parfit 1984; Williams 1970; Lewis 1986, among others) still factor in Locke's core account of personal

[9] On animalism, see Olson (1997 and 2007) and Snowdon (2014).

identity, in terms of the recollection of past experiences or the psychological conscious experience of oneself as a means to ensure that continuity factor needed among multiple possible changes over time: "and as far as this consciousness can be extended backwards to any past Action or Thought, so far reaches the Identity of that Person" (Locke 1975: 27). The fact that a person persists over time does not weigh more heavily than some other facts generally spelled out in either biological or psychological terms, or both. This theory not only leads to inconsistences but also fails to explain personal identity.

Thomas Reid (1710–1796) looked at one of such inconsistencies by considering the example of a person who can now remember her first day in high school but cannot remember her first day in primary school, although, on her first day in high school, she could remember her first day in primary school (Reid 2002: 262). Another inconsistency emerges when considering identity during sleep: Locke's insistence on the necessarily self-aware nature of our thoughts challenges the possibility for self-aware thought during sleep of which we have no recollection (Locke 1975: II, 1, §10). He even argues that our inability to recollect the thoughts we presumably entertained during sleep makes it possible for those thoughts to belong to someone else (1975, II, 1, §11). Joseph Butler ([1736] 1975) raises another problem with Locke's identity theory: it is circular, because memory presupposes personal identity and, consequently, cannot consist in it.[10]

Psychological continuity (in all its possible realizations: memory, personality, projects, preferences) may be regarded as an epistemic criterion for an individual diachronic identity, but it provides neither the necessary nor sufficient conditions for *personal* identity. The question of what it takes for a person to persist over time differs from the question of how to find out whether a person at one time is identical to a person at another time. The epistemic criteria to recognize personal identity over time must not be confused with criteria for identity itself.

A complete knowledge of bodily and psychological properties and their relations would still leave the question of personal identity unanswered. Therefore, personal identity is conceivable in the absence of psychological and bodily relations. An argument points, for example, to changes in body

[10] See also the criticism of John Davis (2003: 5ff., 50–52); Wiggins (2001: 197), Wiggins (1976), and Galen Strawson (2011: Chapter 17) argues against Locke a different notion of circularity.

and psychology (see, e.g., Swinburne in Shoemaker and Swinburne 1984: 22–23). I can conceive myself as having your body and psychology and you as having mine (by the way, more than a thought experiment, it is the experience of thought insertion and delusion of control that proves very common in schizophrenic patients). I could also imagine that I might not have existed, but that instead someone else exists with the same life and body that I actually have. If these scenarios really are metaphysical possibilities, then psychological or bodily relations are neither necessary nor sufficient for personal identity: there is a possible world where I exist without the bodily and psychological properties that I actually have, and another one where the bodily and psychological properties I actually have belong to another person.

Simple Theories of Identity

With all these difficulties in sight, let us explore the simple view. David Wiggins describes identity as "absolute, determinate, and all or nothing, like no other relation but itself"; "irreducible, autonomous and not interestingly supervenient on anything else whatever—like unto no other relation" (2001: 157); "primitive, not a matter of degree" (2001: 188). For Jonathan Lowe, "identity itself is univocal and unanalyzable" (Lowe 2012: 137). He challenges the complex view in the following terms:

> if persons really are fundamental in our ontological scheme, as I very much suspect they are, then we simply should not expect to be able to appeal to other entities of suitable kinds in their case. Rather, persons will be among the things that criteria of identity for other less fundamental kinds of entity appeal to—entities such as mental states, for example. That being so, we should probably conclude that personal identity is primitive and "simple," in the sense that nothing more informative can be said about the identity of persons than that in some cases it just obtains and in others not. (Lowe 2009, ch. 8, 2012: 152)

Lowe regards persons as *individual substances* (Lowe 2012: 155). In a similar vein, Martine Nida-Rümelin affirms that conscious beings "have a non-descriptive individual nature" (2012: 158). She bases this view in a first-person experience (2012: 176):

> Our implicit notion of conscious individuals has its origin in what we are aware of in experience. In any experience we have an implicit awareness of

> ourselves as the subject of the experience, in any action we have an implicit awareness of ourselves as the one who is active in that action, and in any first-person memory we have an implicit awareness of ourselves as persisting across time. We should, or so I suggest, accept a thesis of revelation with respect to what it is to be an experiencing subject: in a non-conceptual and implicit manner we are aware of what it is to be an experiencing subject by being an experiencing subject; we thus have implicit knowledge of what it is to be an experiencing subject given that we are subjects of experience; the nature of what it is to be an experiencing subject is revealed to us by being a subject of experience.[11]

This simple view is connected to Aristotle's notion of identity. For him, "substance has two senses, (A) the ultimate substratum, which is no longer predicated of anything else, and (B) that which, being a 'this', is also separable" (*Metaphysics* V, 8, 1017b 23–56). He also states that "sameness is a unity of the being" (*Metaphysics* V, 9, 1018a 7) and that "The things that are primarily called one are those whose substance is one" (*Metaphysics* V, 6, 1016b 9). That is, substance is a basic notion, and substances have capacities (*dynameis*)—they are powerful particulars. Agents are classes of substances.

Following Joseph Butler's and Thomas Reid's criticisms of John Locke, David S. Oderberg (2005) argues that "there is a vicious circularity in trying to analyze personal identity, as Locke does, in terms of memory or of consciousness in general, since these phenomena presuppose identity (i.e., that it is the *same* person who remembers or is conscious)." This is indeed an "identity problem." However, as Oderberg (2005) notes, it is a problem only to the materialist:

> The "problem of personal identity," as it has come to be known, has a relatively recent currency (due to Locke) and is more fitted to a metaphysical viewpoint that at the very least takes the ideas of disembodied existence and of the immateriality of the soul to be at best highly problematic, at worst not even worth a place in the conceptual landscape on which the problem is grappled with. More strongly, I would venture to say that the problem of personal identity is a problem made for materialists—at least those materialists who take seriously the peculiar ontological status of the mental, the existence of free will and rational agency, and perhaps even the possibility of a future life.

[11] On the simple view, see also Madell (1981, 2015) and Swinburne (2013).

Oderberg defends Aristotelian hylomorphism, a doctrine sustaining that two "co-principles"—"form" and "matter"—are united in a unique substance. He proposes it as an account of personal identity.

Aristotle applies his hylomorphism theory to living beings. For him, the *psyche* (soul) "is the first actuality of a natural organic body" (*On the Soul* II, 1, 412b 4)—namely the form. It provides actuality, identity, and persistence to the living being (cf. Irwin, 1990: 288). As Stephen Brock puts it (2013: 342), it is its "ontological energy." The body, according to Irwin's interpretation of Aristotle (1990: 285; see also 294), is not the "remote" matter—the chemical stuff, i.e., what Descartes considers the body—but the "proximate" matter, which is the ensouled body: the body does not exist outside an ensouled body. Body and soul only exist as a living being—an ensouled body, which is not body or soul or a composite of both, as if they were different entities, but an ontological unity. They are not united accidentally: soul and body are one as form and matter are one, and the actions belong to that unit:

> it is not necessary to ask whether soul and body are one, just as it is not necessary to ask whether the wax and its shape are one, nor generally whether the matter of each thing and that of which it is matter are one. For even if one and being are spoken of in several ways, what is properly so spoken of is the actuality. (*On the Soul* II, 1, 412b 6–9)

Body and soul are one at the substance level, which is the living being, and they are two at the level of substantial principles, but they do not survive apart (cf. *On the Soul* I, 1, 413a 5). In a sense, we have a dualism, but it is not the Cartesian (and probably the emergentist) dualism in which two different substances are accidentally united.

Consequently, the activities of living beings—including human beings—are not of the soul or of the body; they are activities of the whole: "human beings are psychophysical wholes," as William Jaworski puts it (2011: 307). David Bostock (2006: 97) maintains, "the soul is not 'the ghost in the machine' that *makes* the machine work as it does, but is rather the working of the 'machine' itself." Even thinking cannot be done without the body (*On the Soul* I, 1, 403a 10 and 403b 19). Aristotle affirms, "still to say that the soul gets angry is as if one were to say that the soul weaves or builds a house. Probably it is better not to say that the soul pities, or learns, or thinks, but to say rather that the soul is the instrument whereby man does these things" (*On the Soul* I, 4, 408b

13–15). Michael Frede explains, "there is just one subject, the animate object, which, in virtue of the particular kind of form or soul it has, is capable of all these things [actions]" (1995: 97).

While avoiding a dualistic view of the human being, Aristotle's hylomorphic conception of the soul as the form of the body allows for two compatible explanations (*On the Soul* 403a 39 -403b 2):

> the natural philosopher [the scientist] and the logician [philosopher, psychologist] will in every case offer different definitions, e.g., in answer to the question what is anger. The latter will call it a craving for retaliation, or something of the sort; the former will describe it as a surging of the blood and heat around the heart. The one is describing the matter, the other the form or formula of the essence.

The form is closely related to Aristotle's "cause for the sake of which" (later called, final cause). The final cause is the most excellent perfection toward which the form leans for the very reason of being this form. Aristotle emphasizes the roles of the form and the final cause because they provide an ontological root to the unity of the human being.

We might be tempted to consider the soul (only) as the organization of the material constituents of the human being. However, as Frede remarks (1995: 98), Aristotle's proposal goes beyond this. Aristotle himself argues this point (*On the Soul* I, 4). Frede indicates: "we do not try to understand the configuration in terms of the material constituents and their properties, but rather the other way round; we try to understand the material constituents and their properties in terms of the form or organization" (1995: 99). The body is not just matter; it is ensouled matter—"we understand why parts, organs, structure, and other bodily processes of animals are as they are when we understand them in psychic terms" (Irwin 1990: 290): they are teleologically oriented by the form-soul. At the same time, "we correctly believe that psychic states are not reducible to purely material states, because psychic states have a teleological role that resists reduction" (Irwin 1990: 291–292). The form-soul is always the same and explains the persistence of the human person over time.

The ensouled body or incarnated soul is the ground of personal identity. The form does not exist alone but incarnated: this explains why each human person has a unique identity different from the identities of other human persons. The form assures the diachronic identity, and the body, the synchronic identity. However, given that they do not exist separately, the ensouled body—the person—has the diachronic and synchronic identity together.[12]

In addition, all persons build more specific characterizations of themselves over their own ensouled bodies. As Wiggins explains, "The identity of persons coincides (I began by suggesting) with the identity of human beings. Human beings are substances possessed of a specific principle of activity to which, in the course of a life, each one of us gives his own yet more specific, more and more distinctive, determination. Prominent among the specifically human activities is our exercise of the cognitive faculties" (2001: 225–226).

In other words, there is a "first nature" over which a specific human person builds her "second nature." However, as John McDowell asserts, the innate endowment of the first nature "must put limits on the shaping of second nature" (1998: 190). Not everything is possible. The second nature opens a person's eyes to reasons for acting (1998: 189). This is a traditional theme in Aristotle's ethics, and it has also been addressed with its nuances by later philosophers like Kant and Hegel. For Aristotle, habits shape a person's second nature; as he asserts in the *Nicomachean Ethics*, "It makes no small difference whether we form habits of one kind or of another from our very youth; it makes a very great difference, or rather all the difference" (II, 1, 1103b 23–25).

We can apply the metaphysical categories explained in Chapter 3 to the human person. Every human person is a substantial human person, though each one is different from all others. Consider Max, the dog in Chapter 3, as a human person. We can establish that Max is a "substance" (*ousia*) belonging to the *universal* kind "human person." Max is an "instantiation" of the universal kind "human person." Kind is the *essence* of Max, and Max's properties or second nature are its accidents or properties. The essence is necessary, and the accidents are contingent, but this human person, Max, this existent substance is not without its accidents, which—though not necessary—are part of Max's identity.

[12] Oderberg (2007: 66) affirms, "substantial form (or 'form' for short) is a radical or fundamental part of the substance in the sense of constituting it as the kind of substance it is. It is a principle in the sense of being that from which the identity of the substance is derived." There is a discussion about the role of prime matter in constituting identity. I will not delve into this difficult issue.

In my chapter book 2020, I suggested considering John Finnis' conception of personal identity (2005). It comprises the first and second nature, but it also divides the latter into three levels. Finnis bases his identity notion on Aquinas' distinction among four kinds of order of reality in the *Proemium* to his *Commentary on Aristotle's Nicomachean Ethics*. The first order is the natural order. Finnis affirms (2005: 251):

> Each of us, each human person, is by nature an animal with the naturally given capacity not only for all the kinds of activity found in other animals but also for the kinds of understanding, reflection, explanatory judgment, language, and community with other persons [...] In this order and its corresponding mode of explanation, one has one's identity, one's personal identity, as a given (independently of one's thinking about it, or about anything else), as one's subsisting and continuity as this animal.

Finnis looks at the first nature of the human person with her implicit capacities—that is, Aristotle's *zoon echon legon* and *zoon politikon* (rational and political animal). Concerning the second order, Finnis asserts (2005: 251):

> The second kind of order is that of thought itself, of consciousness, observation, attention, consideration, inquiring, understanding, reasoning, and judging. In this order, one has one's identity, one's personal identity [...].

This new level adds the exercise of our rational ability to the basic natural order, leading to a further specification of our identity. For this author, the third kind of order.

> is that which is anticipated and shaped in deliberation, choice, intentional actions, and thus—because choices last in the chooser until negated by a contrary choice—is instantiated in one's subsisting character as self-centered or generous, constant or treacherous, converted or perverted, and so forth. (2005: 251)

This order refers to the person's character or *ethos* which is another specification of its identity. Finally, Finnis writes, "The fourth of the four irreducibly distinct kinds of order is that of culture, or mastery over materials (in the broadest sense of material)—every kind of techne" (2005: 252).

Culture also adds specifications to our personal identity. To sum up, Finnis (2005: 280) quotes from Germain Grisez:

> A person is in all four of the orders, and he embraces all of them in himself. In the person the four orders are distinct, irreducible, yet normally inseparable. The unity of the person is unlike the unity of any entity which is enclosed within one of the four orders. The unity of the person is mysterious and must remain so. This unity is immediately given in human experience, and it cannot be explained discursively, since reason cannot synthesize the distinct orders in a higher positive intelligibility. [...] The unity of the person is not an intelligible principle of a fifth order, distinct from the four, nor is it something like an entity belonging to one or another of the four orders hidden behind all of them. The four aspects of the person all involve and in a way include one another, as the four orders always do. [...] The person is the self who unifies these four distinct and irreducible but normally inseparable aspects. The self [which in the next paragraph Grisez refers of, synonymously, as the soul] is a unifying principle. (1975: 349 and 351)

Finnis concludes (2005: 281):

> In short, there is the identity I have as the particular human being I have been since I was formed in my mother's womb with all the radical capacities, and thus the nature, of a human person. There is the identity I have as a person conscious of much that I have experienced and thought since first I began to be aware of myself (my toes and my fears...) in relation to other things, including people, and began to inquire and assert and deny and argue. There is the identity I have as a person with the character I have made for myself by the choices I have made within the constraints, and between the opportunities, that have arisen from the nature of things, the choices of others, and the accidental conjunctions [...]. And there is the identity I have as an object of cultural forms (a citizen of Australia) and a user of them (English-speaking, tie-wearing) and a projecter of new and partly new personas.

Summarizing, the ground of personal identity is our substance as an ensouled body. This basic identity is "completed" by the exercise of our rationality and social agency, by the habits that shape our character, and by our cultural characteristics.[13] Regardless of all these "components," every person is one and identic. Revisiting Sorokin's trinity, the personality system is influenced by the social and cultural systems.

The Identity of the Economic Agent

The time has come to return to the identity of the economic agent. I have quoted Kirman and Teschl's (2004: 62) characterization: "[t]he economic agent creates, builds, changes, and learns, is self-reflexive and evaluates her actions." These actions are not merely technical—that is, the activity of using means to get certain ends. The self-reflexive character as well as the evaluation capacity and activity pertain to ends. The economic agent, consequently, cannot be a mere maximizer of the use of resources to achieve some given ends. The University of Pittsburg's philosopher Nicholas Rescher deals with economic rationality in Chapter 7 of his book *Rationality* (1988). In Chapter 6, he criticized the Humean notion of rationality and highlighted the relevance of the rationality of ends. A logical consequence follows in his criticism of instrumental maximizing rationality. First, commensurability poses a problem: the idea of utility presupposes its possibility, which is only possible considering utility as as a mere preference index rather than a measure of preferability, thus breaking any connection between maximization and rationality. Rescher states: "When utility no longer reflects what is in someone's real interest, there is no longer any good reason to maintain that maximizing it is 'the rational thing to do'" (1988: 110). Second, we cannot avoid considering the ends and their normative character: "Rational choice is a matter of opting not for what is preferred, but for what is preferable" (1988: 112).

We cannot isolate instrumental rationality from the other motivations of human social actions considered by Weber—value-rational, affective, and traditional motives. I believe that the identity notion shared by Wiggins, Grisez, and Finnis—the simple view of personal identity as a basic substance or personal first nature complemented by specifications of second nature characteristics—fits in very well with the economic agent's demands. Though he tries to keep the different areas of social life—politics, art, science, social life, morality—apart, Schumpeter recognizes that

[13] The narrative (Alasdair MacIntyre, Paul Ricoeur) and normative (Christine Korsgaard) approaches to personal identity (positions that can be classified as complex views of personal identity) consider these characteristics as the core of personal identity.

"while everyone is an economic agent, nobody is solely an economic agent" ([1912] 2002: 135).

However, with this view of personal identity that opens a large room for contingency—each human person is different—one may wonder whether economic science is possible. Indeed, there is a trade-off between the realism of specific cases and the scientific quest for generality. We need to find a balance that depends on the subject and the aims of the investigation. There are grounds for generalizations thanks to the natural conditions and to the common social, psychological, and cultural traits. In fact, new economic currents—like behavioral economics, happiness economics, and institutional economics—are implicitly broadening the conception of the economic agent's identity. We need to overcome the simplistic view of the standard economic maximizing agent. Economics has already begun to do so.

Conclusion

This chapter started by noting Davis' analysis of the orthodox economics conception of the economic agent as atomistic. Concurring with him and with the sociological tradition, I have affirmed that this description does not fit real economic agents. The first step was to explain the characteristics of the economic field in order to subsequently ascertain what conception of the human person's identity matches it. The second step involved defining personal identity, and I chose to embrace a form of its simple view: considering the identity of the human person based on her substantial unity, or the so-called first nature, with a second nature built on top of it by her reason, her social links, and her cultural environment. Finally, I remarked that we need to consider a more substantive and contextual notion of the economic action than the universalist and formal notion of orthodox economics. As Ben Fine states, "possibly more than any other category of social analysis, identity is heavily contextual or situated in its content raising doubts over the applicability of both mathematical models and universal concepts" (2009: 186). Thus, this characterization of the economic agent's identity has relevant implications for the nature of economic sciences.

References

Akerlof, G. and R. Kranton (2000). "Economics and Identity". *Quarterly Journal of Economics CXV*: 715–753.

Aleksandrova, E. V. et al. (2020). "Integralism as a Necessary Direction in Socio-economic Research of Modern Society", *International Journal of Psychosocial Rehabilitation* 24/6: 4992–4999.

Anton Mlinar, I. M. and R. F. Crespo (2021). "Identity Theories in Economics: A phenomenological Approach". In: P. Róna, L. Zsolnai and A. Wincewicz-Price (eds.), *Words, Objects and Events in Economics: The Making of Economic Theory*. Dordrecht: Springer, pp. 193–211.

Aristotle. (1954). *Nicomachean Ethics*. Translated and introduced by Sir David Ross. London: Oxford University Press.

Aristotle. *On the Soul*. Translated by W. S. Hett. Cambridge (Mass.): Harvard University Press.

Bagce, S. and E. Yilmaz (2020). "Critique on Identity Economics," *Journal of Research in Economics* 4/1: 1–28.

Boulu-Reshef, B. (2015) "Towards a Personal Identity Argument to Combine Potentially Conflicting Social Identities," *Review of Social Economy* 73/1: 1–18.

Bostock, D. (2006). *Space, Time, Matter and Form. Essays on Aristotle's Physics*. Oxford: Clarendon Press.

Brock, S. (2013). "Spirituality and Hylemorphism", *Doctor Communis*, Proceedings of the XII Plenary Session, Vatican City, pp. 340–345.

Butler, J. ([1736] 1975). "Of Personal Identity". In: John Perry (ed.) *Personal Identity*. Berkeley and Los Angeles: University of California Press, pp. 99–106.

Crespo, R. F. (2013). *Philosophy of the Economy. An Aristotelian Approach*. Dordrecht: Springer.

Crespo, R. F. (2017). "Aristotelian Hylomorphism: A Framework for Non-Physicalist Philosophers About Philosophy of Mind". In: P. A. Gargiulo and H. L. Mesones (eds.) *Psychiatry and Neuroscience Update. Vol. II: A Translational Approach*. Dordrecht: Springer, pp. 37–46.

Crespo, R. F. (2020). "Identity: Individual and Social, and a Theory of a 'Polymorphic' but Unique Identity". In: W. Dolfsma, D. W. Hands and R. McMaster (eds.), *History, Methodology and Social Economics*. Abingdon: Routledge, pp. 51–66.

Davis, J. B. (2003a). *'The Individual in Economics', Inaugural Lecture*. Faculty of Economics and Econometrics. Amsterdam: Amsterdam University Press.

Davis, J. B. (2003b). *The Theory of the Individual in Economics*. London and New York: Routledge.

Davis, J. B. (2011). *Individuals and Identity in Economics*. Cambridge: Cambridge University Press.

Davis, J. B. (2021). "Introduction: The Status of the Concept of Identity in Economics," *Forum for Social Economy* 50/1: 1–9.

Fine, B. (2009). "The Economics of Identity and the Identity of Economics?," *Cambridge Journal of Economics* 33/2: 175–191.

Finnis, J. M. (2005). "'The Thing I am': Personal Identity in Aquinas and Shakespeare", *Social Philosophy and Policy* 22/2: 250–282.

Frede, M. (1995). "On Aristotle's Conception of the Soul". In: M. Nussbaum and A. Oksenberg Rorty (eds.), *Essays on Aristotle's De Anima.* Oxford: Oxford University Press, pp. 93–107.

Frey, B. and A. Briviba (2021). "Two Types of Cultural Economics: Institutions and Norms". https://papers.ssrn.com/sol3/papers.cfm?abstract_id=3814747. Accessed 20 September 2021.

Gasser, G. and Stefan, M. (2012). Introduction. In: G. Gasser and M. Stefan (eds.) (2012). *Personal Identity: Complex or Simple?* Cambridge: Cambridge University Press, pp. 1–17.

Grisez, G. (1975). *Beyond the New Theism: A Philosophy of Religion.* Notre Dame and London: University of Notre Dame Press.

Irwin, T.W. (1990). *Aristotle's First Principles.* Clarendon Press, Oxford.

Jaworski, W. (2011). *Philosophy of Mind. A Comprehensive Introduction.* Malden and Oxford: Willey-Blackwell.

Kaul, N. (2008). *Imagining Economics Otherwise: Encounters with identity/difference.* Abingdon: Routledge.

Kirman, A. and M. Teschl (2004). "On the Emergence of Economic Identity". *Revue de philosophie économique, 9*, 59–86.

Lewis, D. (1986). *On the Plurality of Worlds.* Oxford: Blackwell.

Locke, J. (1975). *An Essay Concerning Human Understanding*, P. H. Nidditch (ed.). Oxford: Clarendon Press.

Lowe, E. J. (2009). *More Kinds of Being: A Further Study of Individuation, Identity and the Logic of Sortal Terms.* Malden: Blackwell.

Lowe, E. J. (2012). "The Probable Simplicity of Personal Identity". In: G. Gasser and M. Stefan. (eds.) (2012). *Personal Identity: Complex or Simple?* Cambridge: Cambridge University Press, pp. 137–155.

Madell, G. (1981). *The Identity of the Self*. Edinburgh: Edinburgh University Press.

Madell, G. (2015). *The Essence of the Self. In Defense of the Simple View of Personal Identity.* New York, London: Routledge.

McDowell, J. (1998). *Mind, Value, Reality.* Cambridge (Mass.): Harvard University Press.

Mill, J. S. ([1844] 2006). *Essays on Some Unsettled Questions of Political Economy* (Essay V: 'On the Definition of Political Economy; and on the Method of Investigation Proper to It'). In *Collected Works of John Stuart Mill*, Volume 4. Indianapolis: Liberty Fund.

Milonakis, D. and B. Fine (2009). *From Political Economy to Economics*. Abingdon: Routledge.

Mises, L. v. ([1933] 1960). *Epistemological Problems of Economics*. Princeton: D. van Nostrand, (*Grundprobleme der Nationalökonomie*. Jena: Gustav Fisher).

Nida-Rümelin, M. (2012). "The Non-descriptive Individual Nature of Conscious Beings". In: G. Gasser and Stefan, M. (eds.) (2012). *Personal Identity. Complex or Simple?* Cambridge: Cambridge University Press, pp. 157–176.

Nussbaum, M. C. and A. Sen (eds.)(1993). *The Quality of Life*. Oxford: Clarendon Press.

Oderberg, D. S. (2005). "Hylemorphic Dualism". In: E. Frankel Paul, F. D. Miller and J. Paul (eds.) *Personal Identity*. Cambridge: Cambridge University Press, pp. 70–99.

Oderberg, D. S. (2007). *Real Essentialism*. Routledge: New York.

Olson, E. T. (1997). *The Human Animal: Personal Identity without Psychology*. Oxford: Oxford University Press.

Olson, E. T. (2007). *What Are We? A Study in Personal Ontology*. Oxford: Oxford University Press.

Parfit, D. (1984). *Reasons and Persons*. Oxford: Oxford University Press.

Parsons, T. (1951). *The Social System*. Glencoe (Ill.): The Free Press.

Persky, J. (1995). "Retrospectives: The Ethology of *Homo Economicus*", *Journal of Economic Perspectives* 9/2: 221–231.

Reid, T. (2002). *Essays on the Intellectual Powers of Man*, D. R. Brookes (ed.) Pennsylvania: Pennsylvania State University Press.

Rescher, N. (1988). *Rationality: A Philosophical Inquiry into the Nature and the Rationale of Reason*. Oxford: Clarendon Press.

Schumpeter, J. A. ([1912] 2002). "The Economy as a Whole. Seventh Chapter of The Theory of Economic Development," translated by Ursula Backhaus, *Industry and Innovation* 9/1–2: 93–145.

Schumpeter, J. A. (1954). *Economic Doctrine and Method: An Historical Sketch*. New York: Oxford University Press.

Sen, A. (1999). *Reason before Identity*. New Delhi: Oxford University Press.

Shoemaker S. and R. Swinburne. (1984). *Personal Identity*. Oxford: Blackwell.

Snowdon, P. (2014). *Persons, Animals, Ourselves*. Oxford: OUP.

Sorokin, P. A. (1937). *Social and Cultural Dynamics*. Vols. 1, 2, and 3. New York: American Book Company.

Sorokin, P. A. (1941). *Social and Cultural Dynamics*. Volume 4. New York: American Book Company.

Sorokin, P. A. (1947). *Society, Culture, and Personality: Their Structure and Dynamics. A System of General Sociology*. Harper.

Strawson, G. (2011). *Locke on Personal Identity: Consciousness and Concernment*. Princeton: Princeton University Press.

Swinburne, R. (2013). *Mind, Brain, and Free Will.* Oxford: Oxford University Press.
Weber, M. (1949). *Methodology of the Social Sciences.* Glencoe: Free Press.
Weber, M. ([1922] 1978). *Economy and Society*, edited by G. Roth and C. Wittich, Berkeley and Los Angeles: University of California Press (*Wirschaft und Gesellschaft*, 1922).
Wiggins, D. (1976). "Locke, Butler and the Stream of Consciousness: and Men as a Natural Kind", *Philosophy* 51: 131–158.
Wiggins, D. (2001). *Sameness and Substance Renewed.* Cambridge: Cambridge University Press.
Williams, B. (1970). "The Self and the Future", *Philosophical Review* 79: 161–80.
Zuboff, S. (2019). *The Age of Surveillance Capitalism. The Fight for a Human Future at the New Frontier of Power.* New York: Public Affairs.

CHAPTER 6

The Human Person as Worker

In the last chapter, I presented the identity of the economic agent as a real human person with all the pertaining characteristics. The implicit approach to the economic agent in that chapter was as a consumer or an investor, trying to broaden the view of the motivations of her decisions, actions, and reactions. In this chapter, I will deal with the economic agent as a worker. A more complete notion of the identity of the economic agent than the standard economics view of it also bears consequences on the economic agent's desirable characteristics and motivations as a worker.

Work has never been so appreciated as it is in this century. It has not been the job of the higher classes in ancient times until a few decades ago; instead labor has been viewed as the work of slaves, servants, proletarians, or, at best, the bourgeoisie, somehow despised by the high classes.

However, most human beings have always worked. John Dupré and Regenia Gagnier (1996) point out that, historically, the reality of human labor has been examined from workers' subjective perspective until well into the nineteenth century. This standpoint has emphasized either the fatigue and hardships of labor or its importance for human development. This subjective element aside, an objective concern is considered as well: for some, like Aristotle, labor brought about the transformation of a limited nature in order to use what was necessary for the "good life," and no more. Instead, others, like Locke and Calvin (in Max Weber's version),

R. F. Crespo, *The Nature of the Economy*,
https://doi.org/10.1007/978-3-031-02453-5_6

believed the goal of labor was unlimited accumulation.[1] For Weber, work also features an intrinsic relevance for the individual—it is viewed as a fundamental element of individual development.

When neoclassical economics was born, some years after the Industrial Revolution started, labor became a production factor—a commodity with a price—quantified in a salary, which used to be considered in terms of the amount necessary in order for the worker to support himself. If it is not a commodity with a price, like the work of a housewife or a student, it is no longer labor. Labor became one more input with a price, subject to the laws of supply and demand. This view implies a remarkable impoverishment of the notion of work. Its subjective element is cast aside, and the entire value of labor is reduced to its contribution to productive processes and its measurement via a salary (Dupré and Gagnier 1996).[2] Nicholas Grimaldi ends his history of the labor phenomenon by noting that work is ultimately considered only from a financial perspective: the only work that counts is the one that increases capital or produces profit (Grimaldi 2003: 15).

In this light, workers themselves tend to measure the value of their work in terms of the goods that they can buy with its product or its objective value—i.e., the value it produces and not its subjective element. The more they can buy, the more their work is worth, and, as viewed from the *ethos* of neoclassical economics, there is no limit to this "more." Thus, economics provides scientific support for consumerism. As Alasdair MacIntyre (1984: 183) points out, *pleonexia* (greed), a vice in Aristotle's moral view, is now the driving force of modern productive work. People fall prey to this vice even more when possessions bring some sort of prestige. When people's value is measured by their income, their identity is defined by what they can buy with it. The more they have (and show),

[1] Weber's theses in *The Protestant Ethic and the Spirit of Capitalism* (1958) have been heavily debated. However, as Kurt Samuelson asserts (*Religion and Economic Action: the Protestant Work Ethic, the Rise of Capitalism and the Abuses of Scholarship*, University of Toronto Press, 1993), they have been greatly influential.

[2] This article contains several sharp views, but it should be noted that its core argument is Marxist, with all the troublesome implications that entails for an approach to labor (see Hannah Arendt (1959), J. Vialatoux, *Signification humaine du travail*, Les éditions ouvrières, Paris, 1953: 206–210, Rafael Corazón González, *Fundamentos para una filosofía del trabajo*, Cuadernos de Anuario Filosófico, n. 72, Pamplona, 1999: 9–10 and José Ángel García Cuadrado, *Antropología Filosófica*, Pamplona, Eunsa, 2001: 206–207). In this section, I draw from Chapter 9 of my 2013 book.

the more they are worth. The negotiating parties place psychology at the service of earnings and find ways to induce growing consumption pressures on labor through merchandise. This vicious cycle that "rationalizes life" resembles an "iron cage" that traps people, according to Weber and Marx, as Karl Löwith (1982: Chapter 2, *passim*) remarks.

Only when economics incorporated values other than salaries, such as intangible benefits that can be added to the total result, did these values begin to resurface, though *more oeconomico*. This assumes, among other things, an implicit, gross simplification: that these values—actually incommensurable—are exchangeable. For example, an individual may choose to give up his family for a salary, paradoxically hoping to support that family, which is in fact lost in this "exchange." An aspect among others of the individual's identity, her professional calling, becomes sellable. Everything, including people, can be sold through work. Even the healthy notion that some things are priceless and cannot be bought with money has been exploited in credit card advertisements. These considerations are not contrary to the regulated salary system established in the market. Competent authorities' regulation proves necessary to guarantee that salaries cover workers' needs and those of their families. If the resulting cost structure fails to accomplish this goal, the problem lies with the company or the product, not with the worker.

Aristotle did not praise labor. Indeed, he stated that "citizens must not lead the life of mechanics or tradesmen, for such life is ignoble, and inimical to virtue" (*Politics* VII, 9, 1328b 39–40). Yet, it should be noted here that, based on his notions of *praxis* and *poiesis* and the latter's relative relevance, the most important part of labor is its impact on workers rather than its external outcome. From this standpoint, a salary is neither the price of a commodity nor the purpose of labor; rather, it is a condition that makes work possible. Labor's value essentially lies in the fulfillment it affords to workers (and their contribution to the common good of the community where they work—a company—as well as their families and society at large), in terms of the subjective dimension, and in the quality of products or services produced, in objective terms. Labor creates wealth, but therein lies only a part of its economic value.

In the context of the previously described contemporary alienating process, all kinds of reactions begin to emerge, viewing labor—a social and ethical action—as human action, with all its consequences, and recognizing the priority of the subjective aspect of work over its objective aspect. Additionally, this connects labor to human dignity; labor is a good worthy of human beings: with their work, people not only transform nature, adapting it to their needs, but they also accomplish their own fulfillment as persons. Germain Grisez, among other scholars, appreciates work as a basic human good through which people develop their skills and find realization (1983: 124).

Also, the management or business administration discipline, making use of psychological knowledge based on several anthropological views, incorporates several types of motivations for work, far exceeding salaries. Labor allows for the free deployment of human intelligence, creativity, leadership, initiative, collaboration, and directive ability within the framework of workers' individual competent functions. These disciplines are reclaiming part of the wealth lost in the simplification carried out by economics.

Most definitely, the key to an enriching reconsideration of labor lies in recovering its subjective element or *praxis*. Modern economics forwent its practical dimension, leaving work only with its technical dimension. Labor ceased to be a practice endowed with internal goods—an activity in which excellence is a true virtue—to become a merely instrumental task at the service of external goods. As it includes self-realization, labor's subjective or practical dimension necessarily makes reference to the community where people live, given that self-realization cannot be accomplished outside a community. In other words, labor's practical dimension includes its ethical–political element, which is associated with the common good.

Classical philosophy has distinguished between several types of human actions—*facere* (in Latin) or production (*poiesis*), and *agere* or doing (*praxis*)—emphasizing the ethical nature of doing. To fully grasp the notion of labor, its nature, and the priority awarded to its subjective side, it is necessary to understand first that *facere* and *agere* cannot be viewed as absolute, distinct realities—as has often been attempted; rather, they should be regarded as aspects or dimensions of human action. This misunderstanding lies at the root of the often categorical distinctions made between technique and ethics, between morality and business.

In this chapter, I will first explain the meaning of *poiesis-facere* and *praxis-agere* and their consequences for labor. Then, I will introduce the new psychological notions about the desirable treatment of work. Psychological studies claim that, as Amy Wrzesniewski et al. state, "it is important to understand the subjective experience of work: how individuals differ in their experience of the work they do" (1997: 22). In particular, there are different ways of "living" work, the best being to consider work as a "calling." Thus, people have a purpose; they feel called upon to perform a task that bears an impact on others and that is part of an enterprise with a greater reach than their individual possibilities.

Praxis-Agere and *Poiesis-Facere*

Chapter 4 has anticipated the Aristotelian distinction between *praxis* and *poiesis*: "the movement in which the end is present is an action [*praxis*]" (*Metaphysics* IX, 6, 1048b 23 and ss),—seeing, thinking, and living are "immanent," perfect actions or *enêrgeiai*. *Poiesis*, instead, is an act whose result lies outside itself—a transient action; it is also an imperfect action—a movement or *kinesis*. For Aristotle, *práxeis* are moral actions, while *poieseis* are technical. When referring to prudence, Aristotle asserts that it cannot be an art or *techné* because "action (*praxis*) and making (*poiesis*) are different kinds of things. The remaining alternative, then, is that it [prudence] is a true and reasoned state of capacity to act with regard to the things that are good or bad for man. For while making (*poiesis*) has an end other than itself, action cannot; for good action (*eupraxia*) itself is its end" (*NE* VI, 5, 1140b 2–7). When an action is transitive, its objective lies outside, and, thus, it cannot be good or bad. When an action is immanent, however, its objective is the *good* action—*eupraxia*.

This distinction is analytical rather than a distinction among physical actions. When we consider the means to obtain an exterior result, we are dealing with a technique or art, and, when our intent is to perform a good action, we are in the field of *praxis* or practical wisdom. Both considerations refer to the same action: how to make an object and how to make it well, respectively. However, saying that action and production, along with their respective habits, are different acts—at the same time, with the former (action and prudence) linked to morality and the latter (production and art), to technical correction—can lead to confusion, suggesting that only immanent actions can be good and depriving transitive actions of goodness, while disregarding the fact that all transitive actions are ripe with immanency. As a result, culture, technique, and art would have their own individual rules, and their ties to morality would be merely extrinsic.

This division between technique and morality—absent in Greek thinking—underlies contemporary views on technique. For Greek thinkers, as *poiesis* was an expansion of *phýsis* (nature), it made no sense to create controversies between technical and moral realms, given that both are embedded into nature's broader context. There is nothing more contrary to the Aristotelian spirit than a split between technical and ethical rationality.

Aristotle prioritizes the practical dimension, as it is the end of an action, not only its exterior objective. The efficient cause does not act without the prior influence of the final cause (the ends): there is no choice of means without a desired end. Consider this quote from *Nicomachean Ethics*, with some notations between brackets:

> The origin of action—its efficient rather than its final cause—is choice, and that of choice is desire and reasoning with a view to an end [choice is the efficient cause, and the end is the final cause]. This is why choice cannot exist either without reason and intellect or without a moral state [there is no human action outside ethics]; for good action and its opposite cannot exist without a combination of intellect and character. Intellect itself, however, moves nothing, but only the intellect which aims at an end and is practical [without a final cause there is no efficient cause]; for this rules the productive intellect [*poietikês*: technical], as well, since everyone who makes [*poieî*: produce] makes for an end, and that which is made is not an end in the unqualified sense (but only an end in a specific relation, and the end of a specific operation)—only that which is done is that; for good action [*eupraxía*] is an end, and desire aims at this. (1139a 30 a 1139b 4)

As indicated above, for Greek thinkers, *logos* (construed as intellectual understanding, non-calculative thought), *physis* (nature, reality), *praxis* (to act), *poesis* (to make), and *techne* (technique) were perfectly coordinated and interrelated through their orientation toward the *telos*. The *telos* is the end, but it stands at the beginning, fueling the whole process. When *logos* is separated from nature, other *teloi* (if they can still be called by that name) are self-imposed, and the natural order is lost: *logos* becomes solely calculative and reduces everything to a calculation, submitting *physis* (both physical and human nature). We go from being humans to being machines, and we impose an arbitrary order on nature. Thus, obviously, an ecological problem arises.

Thomas Aquinas follows Aristotle closely on the divide separating *praxis* and *poiesis*. He reiterates Aristotelian doctrine numerous times, referring to the *Metaphysics* passage cited earlier.[3] Some Thomistic specifications clearly indicate that what is exempt from morality is not the act but rather its technical element (or dimension), which must be complemented with a corresponding moral judgment. Aquinas sees the radical interference of morality in all human action—be it *agere* or *facere*. He classifies action into human actions—moral and voluntary—and man's acts—instinctive (*Summa Theologiae* I–II, q. 6, Prologue).

While, as already pointed out, Aquinas often cites Aristotle's distinctions between the *praxis*-morality and *poiesis*-technique "pairs," he also clearly indicates the presence of willful intent both in productive actions and contemplative activities. As regards productive action, he claims that the object produced must be good: the product of an art, if not morally good, is not the fruit of this art, but rather opposed to it (I–II, q. 57, a. 3, ad 1). Art, as a habit, he continues, is always oriented toward goodness and virtue—though not perfect virtue, given that moral virtue is necessary for the correct use of art (*In VI Eth*, l. 4, n. 1172). As a result, the technical aspect must be examined by morality, both because of the demands of the goodness of the result and for agents' fulfillment.

Specifically, real action is essentially voluntary and, therefore, moral. Hence, we cannot judge what is fair in any human field based on technique alone. This was very clear in a context like Aquinas'. It is also necessary to keep in mind that Aquinas' historical setting, like Aristotle's, did not feature today's technological advances. For both thinkers, this topic did not require special attention; at present, however, it is necessary to stress that actions involving technique are also moral, and, though in some cases there is nothing to deliberate on, as means are either determined or insignificant details (I–II, q. 14, a. 4c), these actions are human and advance man's fulfillment.

Contemporary moral philosopher Servais Pinckaers (1986: 86, 108) also considers moral and technical elements to be human action dimensions, claiming that a technical dimension can be identified in exterior actions. Yet, he adds, this dimension must be integrated into a new dimension stemming from its relation to the individual undertaking the

[3] He cites *Metaphysics*, IX in, e. g.: I, q. 18, a. 3, ad 1; I–II, q. 3, a.2, ad 3; *In IX Metaph*, l. 2, n. 1788; l. 8, n. 1864; *In VI Eth*, l. 3, n. 1151; *In I Pol*, l. 2, n. 53; *De Potentia*, q. 3, a. 15c. Cfr. also I, q. 23, a.3 ad 1; I, q. 56, a.1c; I-II, q. 31, a. 5c; *In Pol*, Prooemium, n. 6, *De Potentia*, q. 9, a. 9, ad 4; q. 10, a. 1c; *In VI Metaph*, l. 1, n. 1152; *In XI Metaph*, l. 7, n. 2253—all works by Marietti, Taurini.

action and its recipient. In 1976, Karol Wojtyla (1979a: 36–37) stated in the same vein:

> There is an affirmation of Saint Thomas which maintains that man, in his work aimed at the transient dimension which is only made good *secundum quid*: as a worker, doctor, engineer, speaker, writer, etc. However, none of these qualities considered in the immanent dimension of the action, makes man good or bad *simpliciter* as such; it is the ethical nature, that is to say, moral value or disvalue which constitutes objective good or bad of the subject actor and makes man good or bad as a person... In man's actions that have a transient nature, it is essential to seek resolutely that which makes man himself good.

Wojtyla (1979b: 150) views the self-determination of will as the source of the unity of action in its interiority and exteriority. As a result, productive action and its work should be adapted to the perfectible condition of human nature. In culture, including economic, technical, and artistic activities and their results, a moral aspect is necessary, though it may not affect them directly.

There is no action without an end. Ends belong to the agent (and are, as such, intrinsic), though the object of the end might usually be external. This is why all human action is practical above all. Considering any type of action as limited to adapting means to ends is a purely analytical approach; it does not inspire action. Actions begin once their ends are justified and determined, setting out to achieve this end.

An agnostic view on ends leads to a reduction of the theory of action to its technical aspects, canceling out the action itself because there is no efficiency without an end. As humans need to act, an agnostic individual who cannot reason about ends simply determines them through some other human ability (will, emotions, feelings), narrowing reason down to an instrumental role, and, as Aristotle affirms, "clearly it is a mark of much folly not to have one's life regulated with regard to some End" (*Eudemian Ethics* I, 2 1214b 6–12). A consideration of actions that is limited to studying the adaptation of means to ends is tantamount to considering incomplete actions. Means and ends interact and undergo mutual determination. Technical and practical aspects both complement and need each other—the latter having priority over the former. This analysis of human action in its two inseparable dimensions will help re-think labor. As Evandro Agazzi (1992: italics in the original) points out, "a technical activity which ignores this [practical] dimension and restricts its horizons to pure efficiency, forgetting the horizon of duty, will be automatically transformed into a *sub-human activity*."

Understanding labor as a personal human action—as both *agere* and *facere*, *praxis* and *poiesis*, or *actus personae*—bears profound consequences. First, in terms of human action, work should perfect workers. Wojtyla continues (1993: 265–266, italics in the original):

> As I understand St. Thomas' thought, human activity (action) is simultaneously *transitive* and *intransitive*. It is transitive insofar as it tends *beyond the subject*, seeks an expression and effect in the external world, and is objectified in some product. It is intransitive, on the other hand, insofar as it *remains in the* subject, determines the subject's essentially human *fieri*. In acting, we not only perform actions, but we also become ourselves through those actions—we fulfill ourselves in them.

Clearly, the subjective element is the most stable. Human beings need to serve, to do something, to have an impact, as they strive for self-fulfillment. This has been recently highlighted by positive psychology (e.g., Mihaly Csikszentmihalyi 1991; Martin Seligman 2002). This subjective element of the human person is permanent, while the objective element, instead, is contingent and variable. From a subjective point of view, what matters is not so much *what* is done, but *how* it is done. This is what grants dignity to work, which is not just merchandise, a commodity, or an impersonal factor.

Second, with this action, Wojtyla continues, people make external reality more human, building their culture. In this task, it is necessary to reveal the profound relation between human action and truth, goodness, and beauty (1993: 271):

> This [relation] does not take place beyond work, beyond human activity. In fact, the activity also has the dual character of being both transitive and intransitive at the same time, as St. Thomas pointed out. And yet that which is transitive in our culturally creative activity and is expressed externally as an effect, objectification, product, or work can be said to be a result of the particular intensity of that which is intransitive and remains within our disinterested communion with truth, goodness, and beauty.

External outcomes are not the mere result of a product spewed out in mass production. Labor should be but imbued with interiority, lest it is dehumanized and no longer truly human—machines would do it better.

A third consequence is the ethical nature of work. *Agere* and *facere* are interrelated, and it is not possible for workers to be good if they are not ethical. This ethical element includes not only labor itself—with virtues such as honesty, competition, effort to do a good job (objective dimension)—but also an adequate balance with the rest of a person's obligations—family, social, personal (health, rest), and other personal duties.

The priority of the subjective element of labor does not threaten its effectiveness. Recent management literature tends to emphasize the "synergies" among creativity, entrepreneurial spirit, worker satisfaction, and productivity. A long-lasting experience seems to confirm this claim.

Now, even if it were certain that being ethical is profitable (though not always true), being ethical for the sake of profitability is no longer ethical. Schemes based on workers' ownership, management, profit-sharing, and board engagement, as well as shrinking hierarchy pyramids and the distribution of bonus benefits can all be viewed as valuable. However, caution is necessary to prevent that these productivity "rewards" undermine transcendental or intrinsic motivations by placing a price on them. There is an interesting literature on the "crowding-out" effect or motivation corrosion. Bruno Frey (1997: 82–85) explains that intrinsically motivated individuals are denied the opportunity to develop their own interest and engagement in an activity if someone offers them a reward. Shaun Hargreaves Heap (2004: 56) describes this as a version of Gresham's law—"bad money drives good money out of circulation." When a school principal decided to cut down tuition fees for students who were never late, more students began to arrive late, as a price had been put on something priceless. And, thus, parents were no longer motivated to make sure their children got to school on time.

Human beings need to work, both for the above-mentioned anthropological and psychological reasons as well as for more vital reasons—supporting themselves. It may be argued that people could ensure their livelihood in some other way, but labor remains the most dignified way to earn a living. The best way to help people support themselves is giving them the training and tools required to work on their own. Social aid plans—sometimes necessary—must be as temporary as possible, as they undermine human beings' dignity if they persist and become welfare policies. In *Politics* (VI, 5, 1320a 30–1320b 9), Aristotle argues that demagogical policies hinging on distributing money to the poor should be avoided and replaced with policies intended to provide small plots of land or the means necessary to start an agricultural or commercial venture. Indeed, basic economic policies should advance the common good by preventing unemployment. Still, labor goes beyond livelihood: it is a means for people's more complete fulfillment.

In the next section, I will introduce some new concepts developed in the intersection between psychology and management studies that, shedding new light on the motivations of work, underscore the need to rescue the subjective dimension of work.

Why Do I Work?

There is growing interest in the idea of work as being a truly significant experience—that is, revisiting the subjective aspect of work.[4] This has led several scholars to return to an ancient notion with spiritual roots: people's calling. J. Stuart Bunderson and Jeffery A. Thompson (2009) begin their article on the work of zookeepers as follows, "To better understand the nature and characteristics of deeply meaningful work, a small but growing number of management scholars have looked to the notion of work as a personal calling" (Bunderson and Thompson 2009: 3). Indeed, meaningfulness lies at the core of a calling (Dobrow and Tosti-Kharas 2011).

The empirical research on callings that has emerged "suggests that when work provides individuals with opportunities to enact their callings, people tend to see their work as more meaningful because it is experienced as personally fulfilling and having worldly impact" (Rosso et al. 2010: 99). Additionally, organizations are viewed as playing an important role in how employees relate to their work (Rosso et al. 2010).

After the research on *calling* reached its tipping point in 2007 (Duffy and Dik 2013), the academic literature on the topic has flourished, expanding into sub-disciplines such as management and organizational behavior (Duffy and Dik 2013). However, this interest is not entirely new: important precedents have surfaced in recent decades.

[4] For a review of the literature on the meaning of the work, see Rosso et al. (2010). Here I draw from Rodríguez, Mesurado and Crespo (2019).

Though this abundant literature seems to coincide with the construct of calling as a means to imbue work and life with meaning, there is little consensus on how to define this experience (Duffy and Dik 2013). For some, a calling is experienced as a summons originating from outside oneself (Dik and Duffy 2009). For others, it can stem from an internal inclination (Hirschi 2011). Many link it solely to professional work, while others view it as a construct applicable to any area of life (Elangovan et al. 2010). A discussion has also emerged on whether this experience is unique to those who see themselves as religious or can also occur among those who do not adhere to specific beliefs (cf. Hunter et al. 2010; Duffy 2006).

Nonetheless, beyond the numerous differences and varied emphases present in the abundant literature published in recent years, several important agreements have been reached around the definition of callings. According to Amy Wrzesniewski, the two most salient approaches to a calling in modern research view it as springing from inner enjoyment and as a way of contributing to a greater good (Wrzesniewski et al. 2009). The same author refers to these two dimensions in other words by describing people who pursue a calling as those who "love their work and think that it contributes to making the world a better place" (Wrzesniewski et al. 1997: 22). Thus, a calling can be construed as a passion for an activity that is perceived as being socially relevant (Wrzesniewski and Tosti 2006). Shoshana Dobrow (2004) speaks of a sense of passion or a deep enjoyment and satisfaction from engaging in one's work. For Martin Seligman (2002), a calling is linked to the passion for a gratifying activity.

Robert Bellah et al. (1985) posit that when work is embraced as a calling, the job has intrinsic meaning and value, instead of merely being tied to a resulting product or benefit. Certainly, this would indicate that a calling primarily requires that the work involved be intrinsically motivating or, in other words, that the person feel moved to work because the job in itself is interesting, attractive, and somehow satisfying (Amabile et al. 1994). A calling may thus be linked to another construct postulated by positive psychology: optimal experience or flow (Duffy et al. 2011). Flow is defined as an "intense experiential involvement in moment-to-moment activity. Attention is fully invested in the task at hand, and the person functions at his or her fullest capacity" (Csikszentmihalyi et al. 2005: 600). On the other hand, Arnold Bakker's (2008) definition, pertaining to the work context, describes flow at work as "a short-term peak experience at work that is characterized by absorption, work enjoyment and intrinsic work motivation" (2008: 401). The flow construct is regarded as a prototype of intrinsically motivated behavior (Deci and Ryan 2000), and, although it refers to a fluctuating mental state rather than to a stable experience like a calling (Elangovan et al. 2010), it describes what is generally felt during an activity that one is passionate about. Certainly, the individual has an intrinsic state of motivation because the activity produces such a satisfying mental state that the person undertakes the activity without consideration for the external reward and despite the high cost of the energy invested in the task or the efforts required to reach the goal.

The literature on work orientation is mostly based on the distinction between job, career, and calling proposed by Bellah et al. in their book *Habits of the Heart* (1985). For these authors, the way in which people view themselves underlies these different notions of work. When understood as a "job," work is a way to earn money and get by. This reflects a "self" that is defined by success and economic security. When construed as a "professional career," work marks personal advancement throughout life with milestones such as professional success and promotions. In this case, the "self" is defined by a broader type of success that includes social status and prestige as well as a sense of growth in power and skill, turning work into a source of self-esteem. More importantly, when understood as a "calling," a person's work is morally inseparable from her life. It is an activity that has meaning and value in itself and goes beyond the product or profit resulting from it.

Bellah et al. emphasize that people's callings connect them with their co-workers and the general community because they see their work as a contribution to the common good. In this sense, these authors maintain that a distinctive motivation underlying a calling is, as mentioned, "to make the world a better place" (Bellah et al. 1985: 102). This expression was coined to describe this construct in the literature that characterizes it. "Having a calling orientation was linked to several self-reported benefits, including higher life, health, and job satisfaction, and lower absenteeism than job- and career-oriented respondents" (Berg et al. 2010: 974).

Wrzesniewski et al. (1997) were the first to confirm that the distinction made by Bellah among job, career, and calling could be of interest and potentially applied in organizational psychology. Wrzesniewski et al. (1997) posit that for work activities viewed as an end in themselves to be understood as callings, they must also be perceived as socially valuable. The authors of the *Work-Life Questionnaire* also make use of the expression "to make the world a better place" to characterize a *calling* (Wrzesniewski et al. 1997). The phrase is quoted by various authors in relation to this construct, including Grant (2007), Bunderson and Thompson (2009), Elangovan et al. (2010). This consensus suggests that perceiving one's work as socially relevant is a necessary condition for that work to qualify as a calling. Those who work in pursuit of a calling to change the world and do not see this impact become frustrated. However, as Bryan Dik and Ryan Duffy state, "some people help others directly and tangibly—teachers, social workers, and physicians, for example. Others do so indirectly, but not insignificantly" (Dik and Duffy 2012: 13). Indeed, teachers directly experience the impact of their classes as they see their students learning. But what happens, for example, to those individuals who work in marketing or finance at a multinational corporation? In these cases, the perception of the social relevance of their work is mediated by an organization. The organization itself is socially relevant, and thus it transforms the work of every employee into something valuable for society, through its integration into the company's mission.

Nonetheless, as M. Teresa Cardador suggests, "despite a growing interest in callings in the academic and popular press, there is a paucity of empirical research examining the workplace implications of callings" (Cardador et al. 2011: 1). In addition, although it is acknowledged that people can experience callings in any activity (Dik and Duffy 2009; Wrzesniewski et al. 1997), up to now, most studies on this construct have been centered on jobs or organizations with an explicit social purpose (teachers, health workers, public sector employees, zookeepers, etc.). For this reason, in Rodríguez, Mesurado, and Crespo (2019), we have sought to study the conceptualization of callings among employees at multinational corporations, which employ a significant share of workers around the world and where work can seem more oriented toward profitability rather than providing a relevant service to society. We chose one of the world's largest technology companies—one that represents the most emblematic context of modern work. We believe that this study has provided some clues as to what organizations can do to facilitate the calling experience among their employees. The results of this study suggest that the distinction among job, career, and calling also exists in the corporate world.

This research also confirms that there is a close relationship between calling and *flow*. Although, as noted earlier, there were theoretical justifications that presupposed a relationship between calling and flow, as far as we know, this is the first study that has empirically demonstrated this relationship. We can affirm that those individuals who work in pursuit of a calling develop a much stronger connection with the activities that they undertake than individuals who work for money or out of a desire for professional development. Specifically, individuals who follow a calling report higher levels of absorption, enjoyment, and motivation than those who work for professional development and even more so than those who do it for money. For people who pursue a calling, the work in itself is gratifying. Everything sought by doing that particular job is part of the job itself. This statement has important theoretical implications. A person can flow in and out of different types of activities, which means that they can find their calling in different tasks. Mihaly Csikszentmihaly (1991) describes the necessary conditions for a job to produce a flow experience: a balance between challenge and skill, immediate feedback, and clear goals. These three conditions offer an attractive perception of competence (Salanova et al. 2006). Consequently, we conclude that this perception of competence is closely associated with the experience of a calling.

Our aforementioned study (2019) also shows that individuals who identify with a calling in multinational corporations also report that their work has a higher "perceived social relevance" than the work of those who pursue professional development and those who work for money. Hence, if the company is socially valuable and communicates this value, its employees are likely to perceive the work they undertake as valuable to society. Working for a company can thus be gratifying in itself because it implies participating in something that is socially relevant. We conclude that the work that employees perform in a company can be experienced as a calling, but they must perceive their work in the organization as socially relevant, and they need to feel part of its mission (see Dik and Duffy 2012).

These conclusions match the desires of new generations, including millennials: they have a well-established purpose, and they are not willing to commit to a job that does not fit in with that purpose.[5] They do not want to adapt themselves to a job: they want to find a work that

[5] On the drivers of meaningful work among Millennials, see Silvia Soria Sojo (2018). See also Jennifer Deal and Alec Levenson (2016). There is an abundant literature on millennials at work. See, for example, https://www.fond.co/blog/what-millennials-want-in-the-workplace/, https://www.aperianglobal.com/attracting-retaining-millennials-global-workplace/, https://www.kforce.com/articles/millennials-5-ways-were-reinventing-the-workforce/

matches their purpose or calling. Some researchers refer to millennials as the purpose-oriented generation, since they want to have a positive impact in the society. They have a sense of life, and their work must suit it.

Justin Berg et al. (2010: 973) note that people in the United States change jobs 10 times between the ages of 18 and 42, and the same is happening in Europe. They explain this fact by stating that "in addition to these extrinsic benefits [financial rewards and promotions], individuals often seek occupations that will provide fulfillment of core personal values, meaning and purpose, self-expression and opportunities to help others" (2010: 973). People do not look for a job, but for a calling. The role of leaders is to show a purpose that can attract workers and can be adopted as the sense of their life. Referring to millennials, Silvia Soria Sojo asserts (2018: 50):

> They would prefer a job offer where they can find meaning and purpose even if the salary or the growing possibilities on the hierarchy are better on other offers. Together with the evidence that money-oriented millennials are the smallest cohort (8.51%), we can argue that millennials are in the pursuit of purpose at work even if they are struggling to find it, which may lead to enhance their frustration as well as the increase of turnover. Thus, we can see an observable growing global trend that millennials are pursuing purpose at work.

Against this backdrop, the analyses of work carried out by standard economics analyses are terribly simplistic. In addition, when an individual's work is driven by a calling, the effects are not only positive for that person, but also for the performance of her team and of the whole organization (see Wrzesniewski 2003).

https://simonsinek.com/discover/the-secret-to-making-millennials-happy-at-work-isnt-about-millennials/, https://www.wespire.com/10-things-millennials-look-for-in-an-employer/.

Conclusion

This chapter has shown that, given the particularly personal nature of labor, current economics—so concerned with work as a production factor—neglects its most valuable elements almost entirely. These components and their implications have been described at some length, but more attention has been paid to its subjective-immanent-interior and objective-transitive-exterior dimensions, underscoring that they are just dimensions, not distinct actions. As a result, it does not make sense to separate technique and business from ethics, or efficiency concerns from labor-based self-fulfillment. While Aristotle did not consider work as valuable in itself but only as a means to secure the leisure needed for a political life and *eudaimonia*, his ideas on praxis and *poiesis* may be applied to labor in order to revisit its notion within the framework of political economy as a practical science.

The intersection of managerial and psychological studies reveals the advantages of working in pursuit of a calling for the individual, the working group, and the entire corporation. Actually, the so-called millennial generation clearly looks for purpose at work. Organizations also need to genuinely do it if they want to attract and retain talented people. However, the abundant literature on millennials at work often shares the old utilitarian mindset, trying to manipulate instead of motivating them, and this typically backfires. Real motivation calls for organizations' and leaders' authentic, truly embraced values.

All in all, the psychological and managerial literature on work confirms the conclusions of its philosophical analysis: the subjective dimension of work prevails over its objective dimension. Thus, economics should rethink its views on work. The identity of the economic agent—a human person with a plethora of motivations, seeking personal fulfillment and a sense of purpose in life—bears an impact on work that economics cannot fail to consider seriously. Ends do matter… a lot.

References

Agazzi, E. (1992). "Per una riconduzione della razionalità tecnologica entro l' ambito della razionalità pratica", In: Sergio Galvan (a cura di), *Forme di Racionalita Pratica*. Milan: Franco Angeli, pp. 17–39.

Amabile, T. M., K. G. Hill, B. A. Hennessey, and E. M. Tighe. (1994). "The Work Preference Inventory: Assessing Intrinsic and Extrinsic Motivational Orientations", *Journal of Personality and Social Psychology* 66/5: 950–967.
Aquinas, T. (1948). *Summa Theologiae*, Marietti, Taurini/Romae.
Arendt, H. (1959). *The Human Condition*. Chicago: The University of Chicago Press.
Aristotle. Aristoteles Opera, edited by August Immanuel Bekker (1831, 1836, 1870, 1874). Berlin: Academia Regia Borusica.
Aristotle. *Nicomachean Ethics*, translated by Sir David Ross (1925; 1954). London and New York: Oxford University Press.
Aristotle. (1999). *Nicomachean Ethics,* transl. and introduction by Terence Irwin. Indianapolis: Hackett Publishing Company.
Aristotle. *The Basic Works of Aristotle*, edited and with an Introduction by Richard McKeon (1941), Random House (reprint of the translations prepared under the editorship of W. D. Ross, New York: Oxford University Press).
Aristotle. *The Complete Works of Aristotle*, edited by Jonathan Barnes (1984). Princeton: Princeton University Press.
Aristotle. (1958). *The Politics of Aristotle*, edited and translated by Ernest Barker. Oxford: Oxford University Press.
Bakker, A. B. (2008). "The Work-related Flow Inventory: Construction and Initial Validation of the WOLF", *Journal of Vocational Behavior* 72/3: 400–414.
Bellah, R. N., R. Madsen, W. M. Sullivan, A. Swidler, and S. M. Tipton (1985). *Habits of the Heart*: Individualism and Commitment in American life. Berkeley: University of California Press.
Berg, J. M., A. M. Grant and V. Johnson (2010). "When Callings Are Calling: Crafting Work and Leisure in Pursuit of Unanswered Occupational Calling", *Organization Science* 21/5: 973–994.
Bunderson, J. S. and J. A. Thompson (2009). "The Call of the Wild: Zookeepers, Callings, and the Double-edged Sword of Deeply Meaningful Work", *Administrative Science Quarterly* 54/1: 32–57.
Cardador, M. T., E. Dane and M. G. Pratt (2011). "Linking Calling Orientations to Organizational Attachment via Organizational Instrumentality", *Journal of Vocational Behavior* 79/2: 367–378.
Crespo, R. F. (2013). *Philosophy of the Economy: An Aristotelian Approach*. Dordrecht: Springer.
Csikszentmihalyi, M. (1991). *Flow: The Psychology of Optimal Experience*. New York: Harper.
Csikszentmihalyi, M., S. Abuhamdeh, and J. Nakamura. (2005). "Flow". In: A. Elliot (ed.) *Handbook of Competence and Motivation*. New York: Guilford Press, pp. 598–608.

Deal, J. and Levenson, A. (2016). *What Millennials Want From Work*. New York: McGraw-Hill Education.
Deci, E. L. and R. M. Ryan. (2000). "The 'What' and 'Why' of Goal Pursuits: Human Needs and the Self-determination of Behavior", *Psychological Inquiry* 11/4: 227–268.
Dik, B. J. and R. D. Duffy. (2009). "Calling and Vocation at Work Definitions and Prospects for Research and Practice", *The Counseling Psychologist* 37/3: 424–450.
Dik, B. J. and R. D. Duffy. (2012). *Make Your Job a Calling: How the Psychology of Vocation can Change Your Life at Work*. Pennsylvania: Templeton Foundation Press.
Dobrow, S. (2004). "Extreme Subjective Career Success: A New Integrated View of Having a Calling", *Academy of Management Proceedings* 2004/1: B1–B6.
Dobrow, S. R. and J. Tosti-Kharas. (2011). "Calling: The Development of a Scale Measure", *Personnel Psychology* 64/4: 1001–1049.
Duffy, R. D. (2006). "Spirituality, Religion, and Career Development: Current Status and Future Directions", *The Career Development Quarterly* 55/1: 52–63.
Duffy, R. D. and B. J. Dik. (2013). "Research on Calling: What Have We Learned and Where are We Going?" *Journal of Vocational Behavior* 83/3: 428–436.
Duffy, R. D., B. J. Dik, and M. F. Steger. (2011). "Calling and Work-related Outcomes: Career Commitment as a Mediator". *Journal of Vocational Behavior* 78: 210–218.
Dupré, J. and R. Gagnier, 1996. "A Brief History of Work", *Journal of Economic Issues* 30/2: 553–559.
Elangovan, A. R., C. C. Pinder, and M. McLean. (2010). "Callings and Organizational Behavior", *Journal of Vocational Behavior*, 76/3: 428–440.
Frey, B. (1997). *Not Just for Money: An Economic Theory of Personal Motivation*. Aldershot: Elgar.
Grant, A. M. (2007). "Relational Job Design and the Motivation to Make a Prosocial Difference", *Academy of Management Review* 32/2: 393–417.
Grimaldi, N. (2003). "Semántica del trabajo". In: M. Herrero (coord.), *Sociedad del Trabajo y Sociedad del Conocimiento en la Era de la Globalización*. Madrid: Prentice Hall, pp. 11–15.
Grisez, G. (1983) "The Goods that Fulfil Persons". In: G. Grisez, *Christian Moral Principles*. Franciscan Herald Press, pp. 115–140.
Hargreaves Heap, S. (2004). "Economic Rationality". In: J. B. Davis, A. Marciano and J. Runde (eds.), *The Elgar Companion to Economics and Philosophy*. Aldershot: Elgar, pp. 442–60.
Hirschi, A. (2011). "Callings in Career: A Typological Approach to Essential and Optional Components", *Journal of Vocational Behavior* 79/1: 60–73.

Hunter, I., B. J. Dik, and J. H. Banning (2010). "College Students' Perceptions of Calling in Work and Life: A Qualitative Analysis". *Journal of Vocational Behavior* 76/2: 178–186.

Löwith, K. (1982). *Max Weber and Karl Marx*. London: George Allen & Unwin.

MacIntyre, A. (1984). *After Virtue*. Notre Dame, Indiana: University of Notre Dame Press, Second Edition.

Pinckaers, S. (1986). *Ce qu'on ne peut jamais faire*. Fribourg (Suisse): Ed. Un. Fribourg-Ed. du Cerf.

Rodríguez, O., B. Mesurado and R. F. Crespo. (2019). "Calling: Making the World a Better Place from Within Multinational Corporations" *Current Psychology* 38/3: 821–828.

Rosso, B., K. Dekas and A. Wrzesniewski (2010). "On the Meaning of Work: A Theoretical Integration and Review", *Research in Organizational Behavior* 30: 91–127.

Salanova, M., A. B. Bakker and S. Llorens (2006). "Flow at Work: Evidence for an Upward Spiral of Personal and Organizational Resources", *Journal of Happiness Studies* 7/1: 1-22.

Seligman, M. (2002) *Authentic Happiness: Using the New Positive Psychology to Realize Your Potential for Lasting Fulfillment*. New York: Free Press.

Soria Sojo, S. 2018. *The Pursuit of Purpose at Work among Millennials*, Dissertation, Fund. Getulio Vargas, Rio de Janeiro. https://bibliotecadigital.fgv.br/dspace/bitstream/handle/10438/24532/Disserta%C3%A7%C3%A3o%20completa.pdf?sequence=1&isAllowed=y. Accessed 10 October 2021.

Wojtyla, K. (1979a). "Teoria e prassi: un tema umano e cristiano", *Atti del congreso internazionali teoria-prassi*, Genova Barcellona, 1976, Ed. Dominicane, Napoli, vol. 1, pp. 36–37.

Wojtyla, K. (1979b). *The Acting Person*. Dordrecht, Boston and London: D. Reidel Pub. Co.

Wojtyla, K. (1993). "The Problem of the Constitution of Culture Through Human Praxis". In: Karol Wojtyla (ed.) *Person and Community. Selected* Essays, translated by Theresa Sandok, Peter Lang, New York, pp. 263–275 (original: "Il problema del costituirse della cultura attraverso la 'praxis' umana", in *Rivista di Filosofia Neoscolastica*, LXIX, 1977).

Wrzesniewski, A. (2003). "Finding Positive Meaning in Work". In: K. S. Cameron, J. E. Dutton and R. E, Quinn (eds.) *Positive Organizational Scholarship: Foundations of a New Discipline*. San Francisco: Berrett-Koehler Publishers, pp. 296–308.

Wrzesniewski, A. and J. Tosti. (2006). "Career as a Calling". *Encyclopedia of Career Development* 1: 71–75

Wrzesniewski, A., K. Dekas and B. Rosso (2009). "Callings". In: S. Lopez and A. Beauchamp (eds.), *Encyclopedia of Positive Psychology*. Oxford, UK: Blackwell, pp. 115–118.

Wrzesniewski, A., C. McCauley, P. Rozin and B. Schwartz (1997). "Jobs, Careers, and Callings: People's Relations to Their Work", *Journal of Research in Personality* 31: 21–33.

CHAPTER 7

The Metaphysics of Social Collectives

Once we have established the metaphysic nature of the economy and of the economic agent as deciding and acting subject, and also as worker, we need to elaborate on the metaphysical nature of the economic system—i.e., the macroeconomy. While this task was postponed in Chapter 4, this chapter will serve as a prelude to it. Macroeconomy refers to the economic dimension of a whole composed of a great number of people interacting between them—a social collective. Thus, before delving into the topic of the nature of the macroeconomy or macroeconomic entities, we have to establish the nature of social collectives. This task is usually called "social ontology," but, according to the definitions of metaphysics and ontology—the latter dealing with an inventory of what exists, and the former, with its nature—adopted in Chapter 2, this chapter will explore the metaphysics of social collectives.

Concerning the ontology or inventory of social collectives, the "cast" is very large. In an attempt to cover a wide range of social collectives, Brian Epstein (2015: 133–136) distinguishes four "paradigms of groups": crowds, masses, formal organizations, and small groups. I use the term "collective" because I think it encompasses different forms of social "ensembles" (an unusual term to refer to them).[1]

[1] Margaret Gilbert (2001: 109, nt. 1) uses the term "collective" because she considers it as the "genus".

R. F. Crespo, *The Nature of the Economy*,
https://doi.org/10.1007/978-3-031-02453-5_7

From Plato and the Sophists to the present, philosophers have asked themselves about the nature of "social collectives." Hand in hand with the emergence of sociological studies, philosophical reflections on this topic have flared up in the twentieth century and the current century, leading to an increasingly sophisticated analysis.[2] Upon describing social ontology, Epstein initially states that "Social ontology is the study of the nature and properties of the social world. It is concerned with analyzing the various entities in the world that arise from social interaction" (2018: 1).

Then, he introduces a core topic in social ontology:

> A prominent topic in social ontology is the analysis of social groups. Do social groups exist at all? If so, what sorts of entities are they, and how are they created? Is a social group distinct from the collection of people who are its members, and if so, how is it different? What sorts of properties do social groups have? Can they have beliefs or intentions? Can they perform actions? And if so, what does it take for a group to believe, intend, or act? (2018: 1)

Deborah Tollefsen (2014: 86) asks similar questions:

> Are social entities as corporations anything "over and above" the individual human beings that comprise them? If we know all the facts about individual psychology, will we understand social processes, relations, and entities? If we went about counting all the things in the world—people, chairs, leaves, snails—would groups be added to the list? Do groups really have attitudes such as belief and intention?

This chapter orbits around these questions. The key or central question around which all other questions revolve is "what sorts of entities are they?," a strictly metaphysical question.

There is a broad variety of views about the nature of social collectives, ranging from extreme "individualist" to extreme "collectivistic" positions. The former maintains that social collectives are nothing more than an aggregate of individuals, whereas the latter holds that collectives are independent of individuals. The former views societies as fictitious,

[2] For a review of issues and positions described by Brian Epstein (2018), see the Stanford Encyclopaedia of Philosophy, entry "Social Ontology": https://plato.stanford.edu/entries/social-ontology/ and https://plato.stanford.edu/entries/social-ontology/history.html.

conceptual, linguistic, or "nominal" entities, while, for the latter, societies are "living" entities, ontologically different from their members, who have intentions, rights, and duties, and perform actions. This last position has also been called "organicism," in the sense that a social collective resembles a living organism, with its members viewed only as parts of the whole that not only have intentions and feelings but also perform actions.

Before moving forward, we need to make two distinctions. First, we must look at the methodological and ontological positions concerning societies. Ontological individualists often uphold methodological individualism, while ontological holists (or collectivists) advocate methodological holism (collectivism), albeit also admitting methodological individualist explanations (see Julie Zahle 2021). Here I deal with the ontological theories. Deborah Tollefsen (2017: 391) explains:

> Ontological holism is the view that there are social entities that exist and these entities are something over and above a mere aggregate of individuals. Ontological individualism is the view that there are no non-reducible social entities—only individuals exist. Methodological holism and individualism are claims about the proper methodology of the social sciences. Although ontological commitments often motivate methodological ones [...] they need not. One could agree that only individuals exist but deny that all social facts can be explained solely in terms of individual psychological states. Likewise, one could argue that groups exist but provide an explanation of them solely in terms of individual psychological states and the relations between them. That is, one could be an explanatory reductionist without being an ontological eliminativist.

Originally posited by Philip Pettit (1996: 117ff.), the second distinction separates the two pairs individualism-collectivism and atomism-holism, which are often mixed up (like in both previous paragraphs). Tollefsen elaborates (2017: 391):

> According to atomism, it is possible for human beings to develop all the capacities characteristic of human beings in complete isolation from other humans. There is no incoherence, for the atomist, in the possibility of a solitary individual. [...] The opposite of atomism is often called holism, not to be confused with methodological holism. Holism is the view that human agents depend non-causally on their social relations with one another for the possession of distinctive human capacities. They only have these capacities (for instance, the capacity to think) as social beings.

These are ontological views on the nature of human beings. Instead, individualism and collectivism, as explained above, are positions about the nature of social collectives, regarding them as reduced to the individuals in them or viewing collectives as more than their members.

Aristotle had already tacitly considered (and rejected) both extreme versions of these views. Specifically referring to the nature of society and contrary to Plato's monistic notion (*Politics* II, 2), he denies that the *polis* is a substantial unity. He also rejects the sophist concept of the *polis* as a merely accidental unity (*Politics* III, 3). Aristotle's rich metaphysics provides categories that allow for an "intermediate" stance. Concerning the atomism-holist ontological positions, Aristotle defines man as a social animal (*zoon politikon*, *Politics* I, 2 1253a 11), and an isolated man isolated is "either a beast or a god" (*Politics* I, 2 1253a 29).

Additionally, as noted above, there are different kinds of social collectives, and each of them has been defined differently: societies, associations, groups, institutions, communities, corporations, organizations, and so on. Faced with this plethora of definitions, I have decided to adopt a "circumventing strategy" to determine the essence of social collectives—the same strategy used in Chapter 4 for the economy. According to Aristotle, "we ought to use our terms to mean the same things as most people mean by them" (*Topics* II, 2, 110a 16–17). Therefore, I have consulted dictionaries because they provide the meaning of words as used in ordinary language. This review aims to identify characteristics shared by social collectives and to establish whether they have a common essence. This common core will serve as a starting point to ascertain the ontology of social collectives.

Next, after repeating some basic Aristotelian metaphysical notions introduced in Chapters 2 and 3, I will proceed to describe and analyze Aristotle's seminal views on social collectives, which I have found helpful to establish their essence. I am aware of the widespread contemporary debate on the nature of groups, institutions, teams, and so on. However, I prefer to avoid such discussions and approach the topic from this angle in order to gain a "fresh" and useful perspective. In the last section, I will introduce some views that I think have some similarities with Aristotle's and I believe will benefit from Aristotelian notions. This chapter will concentrate on the nature of social collectives regardless of the terminology used.

Dictionary Definitions of Social Collectives

In pursuit of the strategy described above, I have picked the following definitions, italicizing some words that denote a shared characteristic to confer some kind of unity.

Society:	"a large group of people who live together in an organized way, making decisions about how to do things and sharing the work that needs to be done." (https://dictionary.cambridge.org/es-LA/dictionary/english/society.) "A voluntary association of individuals for *common ends*," "an enduring and cooperating social group whose members have developed organized patterns of relationships through *interaction* with one another" (Merriam-Webster).
Association:	"an organization of persons having *a common interest*" (Merriam-Webster). "A group of people who work together in a single organization for *a particular purpose*" (https://dictionary.cambridge.org/dictionary/english/association).
Group:	"a number of people or things that are together in one place" (https://dictionary.cambridge.org/es/diccionario/ingles-espanol/group). "A number of individuals assembled together or having *some unifying relationship*" (Merriam-Webster).
Community:	"the people with *common interests* living in a particular area" (Merriam-Webster). "The people living in one particular area or people who are considered as a unit because of their common interests, social group, or nationality" (https://dictionary.cambridge.org/es/diccionario/ingles/community).
Institution:	"a large and important organization" (https://dictionary.cambridge.org/es/diccionario/ingles-espanol/institution). "An established organization or corporation" (Merriam-Webster).
Organization:	"a group of people who work together for *the same* purpose" (https://dictionary.cambridge.org/es/diccionario/ingles-espanol/organization).

	"An administrative and functional structure" (Merriam-Webster).
Corporation:	"An organization, especially a business, that has a legally separate existence from the people who run it" (https://dictionary.cambridge.org/es-LA/dictionary/english/corporation). "A body formed and authorized by law to act as a single person although constituted by one or more persons and legally endowed with various rights and duties including the capacity of succession" (Merriam-Webster).

These definitions reveal a shared and essential trait of social collectives: a common purpose or interest that unites them and drives interactions among members. With this preliminary conclusion, I will move on to Aristotle's ideas on social collectives. However, I will first recall some elementary Aristotelian metaphysical notions.

Aristotle on Social Metaphysics

I will revisit only two basic categories of Aristotle's metaphysics thinking: the notions of substance and accident as well as his proposed four causes for all entities.

The ways of being of beings or entities—substance and accidents—are explained by Aristotle in his book *Categories*, among other works.He calls them categories or predicates, and there are as many predicates as ways of "existing." Aristotle defines the "primary" entity, substance, as follows: "things are called substances in two ways: a substance is whatever is an ultimate subject, which is no longer said of anything else; and a substance is a this so-and-so which is also separable" (*Metaphysics* V, 8, 1017b 23–5). That is, first, for Aristotle, substances are, by definition, ontologically basic entities, not properties. Second, substance is individual (*a this*) and separable—in the sense that its existence can be affirmed without invoking the existence of anything else. All substances can be identified via identity criteria.

The rest of the entities fall under the heading of accidents. An accident is what happens to a substance either immediately or in a mediated way (through another accident/s). For example, freedom is an accident of the will, which is an accident of human beings; brightness can be an accident of the mind, which is also an accident of human beings. According to

Aristotle, the nine kinds of accidents are quantity, quality, relation, location, time, position, possession, doing (*poieîn* or action), and undergoing (*páschein* or passion).[3]

Aristotle also made a distinction among real causes of entities (*Metaphysics* I, 3–10; *Physics,* II, 3), answering the following questions in order to explain them:

- Efficient cause: who made it?
- Formal cause: why this object or accident and not another? What is it?
- Material cause: what is it made of?
- Final cause: to what end was it made? For Aristotle, the final cause is the cause of causes.

Having found a common feature among social collectives and reintroduced these few basic notions of Aristotle´s metaphysics, I will now introduce his views on the social collectives' nature.

Aristotle begins his *Politics* by stating in the second line of the first paragraph: "all associations aim at some good" (*pasan koinonian agathou tinos stochazontai*; *Politics* I, 1, 1252a 2). In *Politics* III, 6 and 7, he refers to the "common interest (or advantage)" (*koine sympheron*) of cities (*poleis*), noting, for example, that "governments which have a regard for the common interest are constituted in accordance with strict principles of justice [general or legal]" (1279a 17–18).[4] The idea of a shared interest or a common good as the bond that unites people in a society has endured since then, though its content has surely changed.[5] Additionally, as shown in the second section, a common interest is included in the definition of social collectives.

Nonetheless, these different social collectives search for a common good, interest, aim, or purpose in different ways. This search can be natural or designed, and it can be more or less intense or committed. As

[3] Here I follow Aristotle's enumeration of accidents in *Categories* 4, 1b 25–2a 4.

[4] See George Duke (2020: 88) for the translation of *sympheron*.

[5] For example, Adam Smith claims that "[t]he wise and virtuous man is at all times willing that his own private interest should be sacrificed to the public interest of his own particular order or society" (1976: 235 –VI, iii). For John Stuart Mill, a fair government must look for citizens' common good (see Brink 2018).

mentioned in previous chapters, Aristotle frequently uses a logical instrument that helps to gather these different degrees or ways of adherence to a common aim and to consider the search for an end as a common characteristic of social collectives: the analogy of proportion. He applies it to goodness (*Nicomachean Ethics* I, 6 1096a 23) and justice (*Nicomachean Ethics* V, 3 1131a 29, and I have applied it to the economy in Chapter 4). Thus, it is possible to include social collectives that are "weak," because their common purpose or their adherence to it is weak. However, there is a limit to this weakness.[6] Though criticizing Plato's view of the *polis* as a unique substance,[7] Aristotle conceives it as a "strong" aggregate of people. He states:

[6] I think that the difference made by Tony Honoré between a crowd and a group illustrates this limit:

> A group is more than a mere collection of individuals. A crowd is not as such a group, even if it exhibits the pattern of behavior called "crowd behavior." On the other hand, a band of conspirators, a mountaineering expedition, and those engaged in a scientific experiment form groups. What unites them? *A common purpose or activity is no doubt necessary*, but hardly seems sufficient... Some shared understanding as to the means to be pursued is also needed. To constitute a group there must be an element of prescription, which limits the freedom of at least some of the group simply to go their own way. (1987: 34, my cursive)

Honoré mentions some consequences of having a common purpose or activity—I will come back to them later. At this point, I want to highlight that a mere crowd is not a social collective because its common aim is merely circumstantial. Carl Welleman (1995), following Honoré's criterion, considers other groupings that are not "social collectives":

> Very different from organizations are unorganized groups. Among the latter, it is important to distinguish between collections and classes, between groups of people gathered together at some place at some time and groups of people classified together on the basis of a common attribute. (1995: 169)
>
> The togetherness of this sort of group [collections] is purely spatiotemporal. The several members of a collection belong to the same group because they are at the same place at the same time. (1995: 169)

The unity of any class consists simply of some common attribute possessed by and some general name predicated of a number of individuals—not the oneness of the ontological or linguistic subject. Hence, statements about a class must be generalizations about its members (1995: 171).

[7] Aristotle states: "For the city is, with regard to its nature, some multitude, and becoming more of a unity it will be a household rather than a city, and a human being rather than a household. For we would say that the household is more of a unity than the city, and the individual more than the household. So even if someone were able to

> It is clear, therefore, that a *polis* is not an association for residence on a common site, or for the sake of preventing mutual injustice and easing exchange. These are indeed conditions which must be present before a polis can exist; but the existence of all these conditions is not enough, in itself, to constitute a polis. What constitutes a polis is an association of households and clans in a good life (*eû zên*), for the sake of attaining a perfect and self-sufficing existence (*autárkous*) (...). The end (*télos*) and purpose of a *polis* is the good life, and the institutions of social life are means to that end. (*Politics* III, IX, 1280b 29–35 and 1280b 39– 1281a1)

There is a clear and demanding *telos*: the "good life," which, for Aristotle, is a life of virtues. He even compares the *polis* with a body:

> the city is prior by nature to the household and to each of us, since it is necessary for the whole to be prior to the part. For if the whole [body] is destroyed, there will be no foot or hand, unless homonymously, just as if someone were to say the stone [hand is a hand]; for [a hand] when destroyed will be such [i.e., no better than a stone hand], but everything is defined by its function (or activity) and its capacity [...] So, that the city is both by nature and prior to each [individual] is clear. For if the individual having been separated is not self-sufficient, he will be in a state similar to the other parts in relation to the whole, but he who is not able to be in a community or needs nothing through self-sufficiency is no part of a city, and so is either a beast or a god. (*Politics* I, 2, 1253a3–29)

Upon reading these paragraphs, we may be inclined to think that Aristotle holds an organicist view. However, he also thinks that "not only is the *polis* composed of a number of men: it is also composed of different kinds of men" (*Politics* II, 2 1261a 22–24).[8] Richard Kraut emphasizes the fact that the parts of the body are not uniform and that they have different functions (2002: 274).

Terence W. Irwin states, "any strict analogy [of the city] with other wholes and parts rests on an exaggeration that becomes clear once we examine Aristotle's own arguments" (1990: 405). "This weak organic conception [of the city] does not imply that my whole essence is my

do this, he should not do it; for this would destroy the city" (*Politics* 1261a 16–22; cf. 1263b 29–35).

[8] See Robert Gallagher (2011) analysis on this and related passages of Aristotle and their consequences.

belonging to the city" (Irwin 1990: 420). Along the same lines, Robert Mayhew quotes, "In *Metaphysics* A, he [Aristotle] says that 'the man is free who is [or exists] for the sake of himself and not for the sake of another' (982b 25–26)" (1997: 333). That is, the *polis* is a strong unity, but not a substantial unity that dissolves individuality. As Ernest Barker puts it, "it is not, however, a whole in which the separate existence of the parts is lost: it is, on the contrary, a union of elements which still continue to subsist as parts of the new whole which they form" (1959: 234).

Yet, I have just quoted that "the city is prior by nature to the household and to each of us, since it is necessary for the whole to be prior to the part" (*Politics* I, 2, 1253a 3–4). Indeed, in *Politics* I, 1–2, Aristotle presents two solid metaphysical theses: first, the natural character of the *polis* and, second, the political nature of the human being (*Politics*, I, 2 1253a 2–3). From a metaphysical point of view, it is obvious that, given its substantial nature, the human being takes precedence over the city, which is an association of human beings. Thus, how should Aristotle's statement, "and the *polis* is prior by nature to the house and to each one of us" (1253a 19), be interpreted? Aristotle recognizes the temporal priority of the parts of the *polis* when he explains how a house stems from the union of a man and a woman; a clan stems from the union of many houses, and a *polis* stems from a group of clans. Still, he adds. "for it [the *polis*] is the end of the [former] and the nature is the end" (1252b 31–2). Thus, individuals, houses, and clans have the *polis* as their end, and, in Aristotle's system, the end ("the reason for the sake of which") is the first cause of every reality. As Mayhew affirms, "a human being cannot live without the city because the city is required for the living of a full human life. Humans can perform their highest actions (e.g., virtuous actions, philosophy) only within the city (or after having lived in a city). This is why the city is prior to the individual" (1997: 338).[9] In short, the "raison d'être" of the *polis* is a common end or purpose.

For Aristotle, having a common cause is the condition of all the wholes that are not a mere aggregation of parts: "In the case of all things which have several parts and in which the totality is not, as it were, a mere heap (*soros*), but the whole is something besides the parts, there is a cause" (*Metaphysics* VIII, 6, 1045a 9–10). If the cause is a final cause, the whole is not necessarily a substance, as in the case of social collectives, which he

[9] For the development of an argument applying this idea to modern society, see Charles Taylor (1985).

considers an accidental whole. For Aristotle, groups of things or wholes may be substantial, merely accidental, or unities of order.

We have a substantial unity when, for example, we combine copper and iron, and, as a result, we get bronze: each former substance loses its proper substantiality, and a new substance emerges. This is the strongest whole: the substance. Aristotle affirms, "Now most things are called one from doing or possessing or being affected by or being related to some other thing that is one, but the things that are called one in the primary way are those whose substance is one" (*Metaphysics* V, 6, 1016b 6–9).

We have a merely accidental unity—"a mere heap" (*Metaphysics* VIII, 6, 1045a 9)—when, for example, we randomly put papers one on top of the other, and, as a result, we get a pile of papers. Each paper preserves its substantiality; the unity only adds an accidental property: being above or beneath another paper.

Finally, we have a unity of order when the parts preserve their substantiality and their ends, but the order also allows them to contribute to a common end. The end calls for an order of the parts. Aristotle states that substances cannot consist of substances (*Metaphysics* 1039a 3–4, 1041a 4–5; see Mayhew 1997: 328). In the case of social collectives, given that they are composed of individual substances, they cannot be substantial wholes. The unity stems from the orientation toward an end, which implies an order. This order has a normative character: if we want to attain a goal, we have to act in a specific way (and we need to have authority),[10] and it implies a range of stipulations and conditions depending on the particular purpose of a specific social collective.[11]

[10] Barker states: "As a whole, it [the *polis*] is viewed as composed of parts different in kind, which are subordinates one to another; for in all compounds which form a whole, there may be traced a ruling element and a ruled" (1959: 234). In this sense, he considers the nature of a constitution as vital element (1959: 301).

[11] Aristotle enumerates "all the elements necessary for the existence of the state. Our list of these elements will include what we have called the 'parts' of the state as well as we have termed its 'conditions'. To make such a list we must first determine how many services a state performs; and then we shall easily see how many elements it must contain. The first thing to be provided is food. The next is arts and crafts; for life is a business which needs many tools. The third is arms: the members of a state must bear arms in person, partly in order to maintain authority and repress disobedience, and partly in order to meet any thread of external aggression. The fourth thing which has to be provided is a certain supply of property, alike for domestic use and for military purposes. The fifth (but in order of merit, the first) is an establishment for the service of the gods, or as it is called, public worship. The sixth thing, and the most vitally necessary, is a method

Thomas Aquinas, who commented extensively on Aristotle's work, used the term "unity of order" in reference to society. In his *Commentary on Nicomachean Ethics,* he holds that Aristotle's conception of society as a whole has to be conceived as a whole that possesses not an "absolute unity" but a "unity of order" (*In Ethic*, I, 1, 5; *Contra Gentes*, IV, 35). In *Summa Theologiae,* Aquinas states that "The community is not a *supposit*, an *unum* per se. Thus, its proper effect is orderly relations" (III, 4, a.1, ad 4). In other parts of his *opus,* Aquinas applies this expression to the order of the world, in reference to Aristotle's *Metaphysics* XII, 10. In this passage, Aristotle asserts that all things of the universe are ordered together somehow, but not all alike—they are connected: "for all are ordered together to one end" (1075a 21). The root and nature of the unity of the whole is the end and the consequent order, but it is not a substantial whole. As Nicholas Aroney argues, "he considered all human communities to consist of parts that in some respects have an operation that is independent and in other respects participate in the operations of the whole" (2014: 20).

Indeed, applying this kind of order to the city fits with its nature. As already mentioned, Aristotle denies that the *polis* is a substantial unity, in opposition to Plato's monistic conception of society (*Politics* II, 2). He also denies that the *polis* is a merely accidental unity. In *Politics* III, 3, Aristotle asserts that a city is more than its place, using Babylon to illustrate his point, as this city "had been taken for three days before some part of the inhabitants became aware of the fact" (1276a 29–30).

Ontologically, then, the Aristotelian *polis* is an order—a quality—of relationships—actions of people: an ordered relationship (a *prós ti*). The order stems from the fact that these actions aim at a common goal—that is, a shared thought and intention. This order of relationships among the families that constitute a *polis* is based on the orientation of their actions toward an end. This order is not substantial. Mayhew remarks, "a community can only be a unity in a secondary or derivative manner (as described in *Metaphysics* V 6)" (1997: 331). In addition, the aims of the individuals do not end with the aims of the city. Everyone can have other personal aims that do not conflict with the common ends. As Mayhew

of deciding what is demanded by the public interest and what is just in men's private dealings. These are the services which every state may be said to need" (*Politics* VII, 8, 1328b 3–16).

puts it, "That one can live his life fully only in the city does not mean that one must live a life wholly for the city" (1997: 338).

For John Finnis, this applies to all kinds of social collectives: "a group, in the relevant sense, whether team, club, society, enterprise, corporation, or community, is to be said to exist wherever there is over an appreciable span of time, a co-ordination of activity by a number of persons, in the form of interactions, and with a view to a shared objective" (1980: 153). For him, "sharing of aim rather than multiplicity of interaction is constitutive of human groups, communities, societies" (1980: 152). This shared aim calls for an order. Finnis holds, "The reality of a community is the reality of an order of human, truly personal acts, an order brought into being and maintained by the choices (and dispositions to choose, and responses to choices) of persons" (1989: 271). The aim is not enough to have a social collective, but it is the factor that triggers a concerted ordered or articulated interaction. This interaction depends on the freedom of the people interacting, and thus it is weak. This is the reason why we need authorities and norms.

This order is accidental and adds something to the whole or unity. As Irwin states, "a whole has some further unity beyond the collection of its parts because its identity is distinct from the identity of the collection" (1990: 256).

What is the nature of this new identity—specifically, the social collective? A final cause, which is a common end, triggers the actions—efficient causes—of individuals—material causes—in an orderly fashion—i.e., the formal cause. In sum, the social collective is a collection of ordered relations of people aiming at a common purpose. These are relations between individual persons, and, as such, they are accidents that inhere in them. Are the members of the collectives part of them? Yes, they are, but inasmuch as they are considered in their interactions toward an end. The net of these interactions constitutes a new entity which is also an accident that ultimately inheres in the members of the collective, but it can have its proper actions, as I will argue in the next section.

On the Action of the Whole

Most countries have laws in place establishing the rights and duties of several kinds of social collectives. For example, according to the United Kingdom's "Corporate Manslaughter and Corporate Homicide Act" (2007), "companies and organisations can be found guilty of corporate manslaughter as a result of serious management failures resulting in a gross breach of a duty of care" (https://www.hse.gov.uk/corpmanslaughter/about.htm). In turn, the U.S. "Federal Sentencing Guidelines for Organizations" states that "Organizations, like individuals, can be found guilty of criminal conduct" [...] and that "An entire organization, despite its best efforts to prevent wrongdoing in its ranks, can still be held criminally liable for any of its employees' illegal actions," thus forcing businesses to create and deploy "compliance programs" (https://www.ussc.gov/sites/default/files/pdf/training/organizational-guidelines/ORGOVERVIEW.pdf). There is ample literature on this topic, mostly framed within a legal or moral context and underpinned by an implicit ontology. However, we should not put the cart before the horse: social collectives' moral and legal tenets must stem from their ontological condition. Let us return to Aristotle to analyze this issue from his perspective.

George Duke notes that "Aristotle not only attributes *eudaimonia* but virtues to the *polis* (*Politics* VII.1, 1323b 30–1 and VII.7, 1327b 31–6). This apparent inconsistency can be resolved by allowing that the *polis* sustains predications as a structured unity of individuals with the same purpose, not as a natural whole (*Nicomachean Ethics* I.2, 1094b 2–13; *Politics* VII.7, 1327b 30)" (2020: 105). If we can attribute virtues to the *poleis,* we can also attribute vices to them. Duke then generalizes:

> A part of a whole can act separately from the whole (as a soldier can act separately from the entire army), but the whole nonetheless has an operation that is not proper to its parts, but rather to the whole (e.g. an assault by the entire army). It is therefore justified to ascribe action to the whole, insofar as it is a unity structured by a common goal. (2020: 104)

Indeed, a common good generates a united network of articulated or ordered actions that acts by itself thru its components. It is the soccer team that wins the match; Messi cannot win alone. In fact, players are replaced during the match, but the team wins the game. An orchestra plays a piece of music, and a firm enters into agreements and produces goods. The aim of the collective triggers actions that any individual alone would not perform in the absence of it: there are real relations that modify the individual. This is a plus that, in addition to other pluses, constitutes an accidental network that can act. Also, the social collective produces results that cannot be produced individually.

As a result, we can legitimately speak of the actions of a social collective because of its metaphysical nature, despite its "weakness." Via the articulation of their members, social collectives have aims, perform actions, and produce effects that cannot be achieved individually. In Aristotle's view, accidents—as sets of relationships—may produce other accidents. Ultimately, these aims and accidents inhere in the members through the ordered network of relationships among them.

Some Contemporary Connections

Connecting Aristotle's notions to contemporary ideas can pose a significant risk. We should be careful to avoid "the inveterate use of making Aristotle reason with the categories of the interpreter," as Gianfrancesco Zanetti rightly notes (1993: 20). Aristotle's thoughts on social collectives focus on the *polis*, an institution that does not exist today. In addition, the notion of *telos*, which is relevant to Aristotle, is difficult to find in contemporary thinkers. Still, I think that the differences between Aristotle's and contemporary ideas do not undermine the Aristotelian conception of social ontology explained earlier, which I consider satisfactory and useful, even to enlighten contemporary "analogous" views.

Therefore, as I am writing in the twenty-first century, I will try to briefly highlight some contemporary thinkers' ideas that might have some similarities with Aristotle's views on social ontology. I will look for theories that somewhat take a stance between, despite the anachronism, modern Platonic and Sophist notions—ontological holism and individualism. Unfortunately, the conclusion will not be decisive: it is not possible to establish all the similarities and divergences between Aristotle's idea of collectives and these other thinkers' ideas in just a few pages. This section only intends to mention them, paving the way for further research.

In an often-quoted article, Steven Lukes asserts:

> Society consists of people. Groups consist of people. Institutions consist of people plus rules and roles. Rules are followed (or alternatively not followed) by people and roles are filled by people. Also there are traditions, customs, ideologies, kinship systems, languages: these are ways people act, think and talk. (1968: 120)

He considers these statements as a "truistic social atomism," a statement of "individualistic ontology," as Epstein remarks (2015: 22). We can ambiguously call this "truism" an atomism or individual ontology. However, this does not mean that the social entity is a mere grouping of people: rules, roles, traditions, customs, and so on would not exist outside social collectives, and, consequently, social collectives are more than the mere aggregations of people. These "real things" added entail or presuppose ends and an order, as Aristotle asserts.

Anthony Quinton holds an "intermediate" position between the conception of social collectives as substances and as mere abstractions or fictions. He states:

> Just as I have acknowledged that men are essentially social while rejecting the inference that social objects are on that account more real and concrete than their members, so I agree that social objects are logical constructions while rejecting the inference that they are on that account merely abstractions. (1976: 18)

He regards these kinds of logical constructions as wholes, and their constituting elements are their parts (1976: 19): "social objects are wholes with individual people as their literal parts, as real and as concrete as their members" (1976: 22). This is why, for Quinton, reducing social facts to individual facts is possible but very difficult: we almost always need a reference to a relation between the individual and the social whole (1976: 23–25):

> I believe, then, that all statements about social objects are statements about individuals, their interests, attitudes, decisions and actions. But the predicates of these statements about individuals will essentially, if only implicitly, mention social objects in a way that is not practically or usefully eliminable, even if it is eliminable in principle. (1976: 25)

Quinton states that assigning social collectives' properties to their members is "summative" (1976: 18). As Margaret Gilbert remarks, Quinton "assumes the simple summative account en passant" (2014: 168) and proposes a more complex theory (2014: 184–192).

Gilbert's notion belongs to a number of theories which, albeit differing in some respects, propose diverse forms of "collective intentionality" that could be loosely linked to Aristotle's views. I will include and briefly discuss the social action notions posited by John Searle, Seumas Miller, Gilbert, Raimo Tuomela, and Dave Elder-Vass.

For John Searle, "social ontology is both created by human actions and attitudes but at the same time has an epistemically objective existence and is part of the natural world" (2006: 12). Searle has concentrated his analysis on a particular social collective, the institution. I will not delve into his take on institutions, which has been widely discussed.[12] I only wish to highlight his views on collective intentionality and assignment of function, which I believe can be applied to social collectives in general. Searle uses the expression "collective intentionality" to refer to the collaborative behavior of "two people carrying on a conversation, an orchestra playing a symphony, and two teams playing football [...] I am doing something only as part of our doing something" (2006: 16). He states that "status functions are the glue that holds society together because they create deontic powers, powers that work by creating desire-independent reasons for action" (2006: 12). I would like to underscore three issues here: (1) the social dimension as part of the natural world,[13] (2) the articulation implied by collective intentionality, and (3) the unifying role of functions, as they generate desire-independent reasons for action. Searle holds an individualist perspective, but he considers a plus—collective intentionality and status function—that would not make sense outside a social collective, thus concurring with Lukes.

In my 2016 article, I asserted that Aristotle's and Searle's (e.g., 2005) ideas share the following elements: the idea of a common end underlying Searle's collective intentionality, as well as the normative and unitive character of the reasons that constitute an institution. These elements suppose a teleological perspective, as also noted by Hindriks and Guala (2019).

In their paper, Hindriks and Guala also discuss Seumas Miller's (2001 and 2010) "collective ends" theory. While Miller maintains that his theory is teleological, Hindriks and Guala argue that it is etiological (I will not discuss this here). Miller's 2001 book includes a subtitle that reads, "A Teleological Account." Miller speaks about "joint actions," pointing out (2001: 13):

[12] For discussions on Searle's position, see, among others, Barry Smith (Smith and Searle 2003); Frank Hindriks (2005); J.P Smit, Filip Buekens, and Stan du Plessis (2011); Georgios Papadopoulos (2015), and Francesco Guala (2016a, b).

[13] It should be noted that, for Searle, "natural" means "physical" (see, for example, Searle 1995: Chapter 1).

> joint actions are comprised of individual actions performed to realise a shared end; so my conception is teleological. The end involved in joint actions is a shared or collective end; it is not simply an individual end.

According to Miller, not all joint actions are social actions, which are interpersonal "actions performed in accordance with social forms such as conventions, social norms, institutions, social groups, and the like" (2001: 6). For him, all social actions are performed by individuals: "the 'actions' of organisations are reducible to the actions of individual human persons" (2001: 160). He criticizes Gilbert, stating that she ascribes collective beliefs and intentions to "plural subjects" (2001: 33 and 85–90).

Discussing "joint commitment," Gilbert asserts, "in the process of joint commitment, two or more people jointly commit the same two or more people" (2014: 7). It is not a "singularist" commitment made by two individuals but a "joint" commitment. Gilbert has considered that jointly-committed people's collective preferences "are not reducible to a set of correlative individual preferences" (2001: 114). She states:

> A joint commitment is not a conjunction of personal commitments [...] a joint commitment is a commitment to do something as a body, where "do" is interpreted broadly to include beliefs and preference among other things [...] I say that those who are jointly committed to do something as a body constitute a *plural subject*. (2001: 115)

For me, Epstein interprets Gilbert correctly when he affirms that she sees joint commitment to a goal as different from individual commitments to the same goal (2015: 252). However, Gilbert clarifies:

> it is worth emphasizing, then, that I have never intended to suggest that there is any collective or group consciousness that is somehow independent of the consciousness of any individual group member. Rather, I see joint commitment as a precondition of the correct ascription of, say, a particular attitude, to a given population of persons. Then we—a plurality of persons—are the subject of that attitude. In other terms, it is ours. (2014: 9)

Though it may seem that her notion of joint commitment refers to small groups, in her 2006 book, Gilbert applies it to societies. Nevertheless, she does not provide a definition of social collectives.

Raimo Tuomela sustains that the position he advocates in his book *Social Ontology: Collective Intentionality and Group Agents* is "a weak conceptual and epistemic collectivism" that may be seen as "a common-sense alternative that lies somewhere between the extreme group-centeredness of German idealism and the conceptually impoverished framework of rational choice theory as we now have it" (2013: 4).[14] The book subtitle mentions two central elements of his proposal. Let us briefly explore the meaning of these two notions.

For Tuomela, "collective intentionality" (or the "we-mode" framework) is "the cement of society" (2013: x). It is a necessary complement of the individual "I-mode" to explain the existence and behavior of social groups. From his standpoint, "people are generally able to engage in both we-mode and I-mode thinking and acting, but the we-mode is not reducible to the I-mode" (2013: 93). The we-mode is not a sum of I-modes.

According to this author, a central driver for the emergence of a social group is to have a reason for action (2013: 11, 34, 38) or an *ethos* (2013: 16, 26, 30, 34). This reason or *ethos* triggers the need to order functions and actions (2013: 1, 2, 8, 16, 44) and entails the authoritative character of them (2013: 9, 34). He states: "we-mode action is based on a group reason and is thus performed for the reason of promoting the group's interests" (2013: 38).

What is Tuomela's ontological definition of social groups? Though he has called it "weak," he thinks that the we-mode is a "conceptually and ontologically strong notion" (2013: 53). He affirms:

> An organized social group [...] exists as an interactive social system (a kind of social whole) that consists of interrelated individuals such that this system is, through them, capable of producing uniform actions and outcomes. This unity arises through the members' bringing about these actions and outcomes jointly, as a group. This kind of group organized as a social system is generally not conceptually reducible [...] to the properties and relationships between the members [...] This irreducibility is in part due to the emergent features that an organized social system (group) has because of its members' interdependencies and interaction, as well as changing membership in the group. (2013: 21–22)

[14] Tuomela notes that some elements drawn from Jean-Jacques Rousseau's, William McDougall's, Alfred Vierkandt's, and Ferdinand Tönnies's views are relevant for his theory (2013: 93–94).

There is an emergent plus, which is the system of interdependencies and interactions: Tuomela thinks that social unity cannot be explained by supervenience (which implies parallel properties and changes in the micro- and macro-levels), but by emergence (2013: 16, 21, 53, 91).[15] However, this does not imply that he sees the social group as a substance (in Aristotelian words) or an object. This is made clear when he characterizes "group agency," the second notion in the subtitle of his book. He states:

> A group agent in the sense of this book is not an *intrinsically* intentional agent with raw feels and qualia, as contrasted with ordinary embodied human agents. The functional and intentional existence of the group is *extrinsic*. (2013: 3, italics in the text)

> The only causally initiating agentive motors are the individual agents, and hence the agency of group agents must ontologically bottom out in the behavior of its members. The group may nevertheless enjoy an autonomy that makes reduction to the private intentional psychologies of its members unfeasible, if not impossible. (2013: 22)

> A group agent is constituted by a (sometimes emergent) collection of interdependent and interacting individuals, and it acts as a group in virtue of its members' actions". (2013: 51)

In turn, Tuomela also argues that "functional" group agents (that can even include states—2013: 21) do have many functions:

> we can on functional grounds attribute as-if mental states such as wants, intentions, and beliefs, as well as actions and responsibility to these groups. Such group agents are not intrinsically intentional agents ("persons") comparable to human beings, but they can on functional and epistemic as well as practical grounds be viewed and accepted as extrinsically intentional agents with attributed quasi-mental properties. (2013: x)

[15] He criticizes List and Pettit (2011) for this reason (2013: 53–54). This section does not intend to discuss the different versions of supervenience and emergence in the specific literature.

I find these different proposals of collective intentionality truly enlightening, albeit probably differing in ontological commitments about the ontological nature of social collectives and their supposed intentions and actions. At the same time, I think that they would benefit from taking into account Aristotle's metaphysical notions.[16]

These positions about social collectives include an extra interrelationality. Using the metaphysical categories employed by Epstein (2015: Chapter 5), they all implicitly deal with an ontology of individual facts plus "properties." Broadly considered in line with Aristotle's accidents, properties are characteristics of "objects": "Objects actually have (or instantiate) some properties, and possibly have others" (Epstein 2015: 64). Relations between properties can be established: "the proposition Bill is taller than Hillary consists of two objects—Bill and Hillary—and a binary relation that holds between them—the 'is taller than' relation" (Epstein 2015: 64).

Thus, regardless of the limitations of comparing two metaphysical sets of categories, we can state that, for Aristotle, a social collective is also a set of relations (a kind of property) between people (objects) ordered toward achieving a common goal.

Finally, "critical realists" like Roy Bhaskar and Dave Elder-Vass resort to the concept of emergence. Elder-Vass defends a "relational theory of emergence" and the ascription of causal powers to social entities (2010: 14). In his opinion, there are wholes—structures—that consist of "the set of relations between the parts that only exist when this structure is present" (2014: 45). These relations cause interactions that produce causal effects. This author states that "entities at each level can have emergent causal powers: powers to affect the world that would not be possessed by the parts if they were not organised into entities of this nature" (2014: 45). Applied to social collectives, Elder-Vass points out that "As a result of its members being committed to interact in the ways specified in their roles, the organisation has the capacity to have a causal impact on the world that its members would not have if they were not parts of the organisation concerned" (2014: 49). He (2014) also argues that social entities are often composed of both human agents and

[16] Epstein is critical of Gilbert's and Tuomela's theories (2015: Chapter 17) on the basis that they are only applicable to small groups. This is not, as I have mentioned, Gilbert's and Tuomela's intention.

non-human material objects, and that both may make essential contributions to their causal influence. This position bears some similarities with Aristotelian theories, though this is to be taken with some reservation.

Conclusion

For Aristotle, social collectives consist of ordered interactions of substantial individuals aiming at common ends. These collectives cannot be reduced to the individuals who orderly interact, because their ends and interactions, though inhering in the individuals, are ontologically different from them. The end is a real cause of substance; relations and actions are real accidents with ontological density, regardless of their inherence in substances. Real relations, though not substances, exist independently of being or not known and add something real. A social relationship adds an order toward an end that proves instrumental to explain individual social actions. This is why, for Aristotle, social collectives are more than the sum of their individual members.

Social collectives have a unity of order, which is weaker than the substantial unity of ontological holism. Was Aristotle an ontological individualist or an ontological holist? He was neither. It would be anachronic to strictly assign Aristotle's thoughts on the nature of societies to these categories.

Finally, I have described some contemporary "intermediate positions" on social collectives. I have claimed that these views would probably benefit from considering Aristotle's related ideas because his metaphysical categories pave the way to a notion of social collectives that is neither substantialist nor merely fictional. While not entirely free of tension, this seems consistent with these contemporary authors' perspective on the subject. However, as anticipated, this claim should be regarded only as a suggestion that requires further analysis.

With these developments in mind, it is now time to take a look at the nature of macroeconomic entities and their ties to microeconomic entities.

References

Aquinas. (1964). *In Decem Libros Ethicorum Aristotelis ad Nicomachum Expositio*. Torino: Marietti.

Aquinas. *Summa Theologica.* http://www.newadvent.org/summa/. Accessed 27 May 2019.

Aquinas (1922.). *Summa contra Gentiles.* Torino: Marietti.

Aristotle. (1954). *Nicomachean Ethics.* Translated and Introduced by Sir David Ross. Oxford: Oxford University Press.

Aristotle. *Nicomachean Ethics.* Translated by Terence Irwin (second edition 1999). Indianapolis and Cambridge: Hackett.

Aristotle. (1958). *Politics.* Edited and Translated by Ernest Barker. Oxford: Oxford University Press.

Aristotle. (1943). *Politics.* Translated by Benjamin Jowett. New York: Random House.

Aristotle. *The Complete Works of Aristotle.* Edited by Jonathan Barnes (1984). Princeton: Princeton University Press.

Aroney, N. (2014). "Subsidiarity in the Writings of Aristotle and Aquinas". In: M. Evans and A. Zimmermann (eds.) *Global Perspectives on Subsidiarity.* Dordrecht: Springer, pp. 9–27.

Barker, E. (1959). *The Political Thought of Plato and Aristotle.* New York: Dover Publications.

Brink, D. (2018). "Mill's Moral and Political Philosophy", *The Stanford Encyclopedia of Philosophy* (Winter 2018 Edition), Edward N. Zalta (ed.), https://plato.stanford.edu/archives/win2018/entries/mill-moral-political/. Accessed 20 May 2020.

Duke, G. (2020). *Aristotle and Law: The Politics of* Nomos. Cambridge: Cambridge University Press.

Elder-Vass, D. (2010). *The Causal Power of Social Structures. Emergence, Structure and Agency.* Cambridge: Cambridge University Press.

Elder-Vass, D. (2014). "Social Entities and the Basis of Their Power". In: J. Zahle and F. Collin (eds.) *Rethinking the Individualism-Holism Debate. Essays in the Philosophy of Social Science.* Dordrecht: Springer, pp. 39–54.

Elder-Vass, D. (2017). "Material Parts in Social Structures", *Journal of Social Ontology* 3/1, pp. 1–17.

Epstein, B. (2015). *The Ant Trap: Rebuilding the Foundations of the Social Sciences.* Oxford: Oxford University Press.

Epstein, B. (2018). "Social Ontology", *The Stanford Encyclopedia of Philosophy* (Summer 2018 Edition), Edward N. Zalta (ed.), https://plato.stanford.edu/archives/sum2018/entries/social-ontology/. Accessed 10 May 2020.

Gallagher, R. L. (2011). "Aristotle on *Eidei Diapherontoi*", *British Journal for the History of Philosophy* 19/3: 363–384.

Finnis, J. M. (1980). *Natural Law and Natural Rights.* Oxford: Clarendon Press.

Finnis, J. M. (1989). "Persons and Their Associations", *Proceedings of the Aristotelian Society* Suppl. Vol. LXIII, pp. 267–274.

Gilbert, M. (2001). "Collective Preferences, Obligations, and Rational Choice", *Economics and Philosophy* 17: 109–119.
Gilbert, M. (2006). *A Theory of Political Obligation: Membership, Commitment, and the Bonds of Society*. Oxford: Oxford University Press.
Gilbert, M. (2014). *Joint Commitment: How We Make the Social World*. Oxford: Oxford University Press.
Guala, F. (2016a). "Philosophy of the Social Sciences: Naturalism and Anti-naturalism". In: P. Humphreys (ed.) *The Oxford Handbook of Philosophy of Science*. Oxford University Press, pp. 43–64.
Guala, F. (2016b). *Understanding Institutions: The Science and Philosophy of Living Together*. Princeton University Press: Princeton.
Hindriks, F. (2005). *Rules & Institutions: Essays on Meaning, Speech Acts and Social Ontology*. Doctoral Thesis, Erasmus Universiteit Rotterdam. Amsterdam: Kloof Booksellers & Scientia Verlag.
Hindriks, F. and F. Guala (2019). "The Functions of Institutions: Etiology and Teleology", *Synthese*, online first.
Honoré, T. (1987). *Making Law Bind*. Oxford: Clarendon Press.
Irwin, T. W. (1990). *Aristotle's First Principles*. Oxford: Clarendon Press.
Kraut, R. (2002). *Aristotle: Political Philosophy*. Oxford: Oxford University Press.
List, C. and P. Pettit (2011). *Group Agency: The Possibility, Design and Status of Corporate Agents*. New York: Oxford University Press.
Lukes, S. (1968). "Methodological Individualism Reconsidered", *The British Journal of Sociology* 19/2: 119–129.
Mayhew, R. (1997). "Part and Whole in Aristotle's Political Philosophy", *The Journal of Ethics* 1/4: 325–340.
Miller, S. (2001). *Social Action: A Teleological Account*. Cambridge: Cambridge University Press.
Miller, S. (2010). *The Moral Foundations of Social Institutions*. Cambridge: Cambridge University Press.
Papadopoulos, G. (2015). "Collective Intentionality and the State Theory of Money", *Erasmus Journal for Philosophy and Economics*, 8/2: 1–20.
Pettit, P. (1996). *The Common Mind: An Essay on Psychology, Society, and Politics*. New York and Oxford: Oxford University Press.
Quinton, A. (1976). "Social Objects", *Proceedings of the Aristotelian Society, New Series*, 76 (1975–1976): 1–27+viii.
Searle, J. R. (1995). *The Construction of Social Reality*. New York: Free Press.
Searle, J. R. (2005). "What is An institution?", *Journal of Institutional Economics* 1/1: 1–22.
Searle, J. R. (2006). "Social ontology: Some basic principles", *Anthropological Theory* 6: 12–29.
Smit, J. P., F. Buekens and S. du Plessis (2011). "What Is Money? An Alternative to Searle's Institutional Facts", *Economics and Philosophy* 27: 1–22.

Smith, A. (1976). *The Theory of Moral Sentiments*, Glasgow Edition. Oxford: Oxford University Press.

Smith, B. and J. Searle (2003). "The Construction of Social Reality: An Exchange", *American Journal of Economics and Sociology* 62: 285–309.

Taylor, C. (1985). "Atomism". In: Ch. Taylor (ed.) *Philosophy and the Human Sciences. Philosophical papers 2.* Cambridge: Cambridge University Press, pp. 187–210.

Tollefsen, D. (2014). "Social Ontology". In: N. Cartwright and E. Montuschi (eds.) *Philosophy of Social Science: A New Introduction.* Oxford: Oxford University Press, pp. 85–101.

Tollefsen, D. (2017). "Collective Intentionality and Methodology in the Social Sciences". In: M. Jankovic and K. Ludwig (eds.) *The Routledge Handbook of Collective Intentionality*, pp. 389–401.

Tuomela, R. (2013). *Social Ontology. Collective Intentionality and Group Agents.* Oxford: Oxford University Press.

Welleman, C. (1995). *Real Rights.* New York and Oxford: Oxford University Press.

Wiggins, D. (2001). *Substance and Sameness Renewed.* Oxford: Oxford University Press.

Zahle, J. (2021). "Limits to levels in the methodological individualism–holism debate", *Synthese* 198: 6435–6454.

Zanetti, G. (1993). *La nozione di Giustizia in Aristotele.* Bologna: Il Mulino.

CHAPTER 8

On the Relation Between Micro- and Macroeconomic "Entities": A Philosophical Approach

This whole book hinges on the premise that philosophy is useful, though not always so considered. I will argue that this premise applies to the relation between micro- and macroeconomic entities. Every era has had a worldview, and today's is physicalist. "Physicalism is the thesis that everything is physical" (Stoljar 2015: 1; thus, it is a reductionist thesis). As affirmed in the Introduction of this book, this contemporary reductionist *ethos* has infiltrated the field of economics. Jerry Fodor's (1974) famous article "Special Sciences," which defends psychology from reductionist projects, is an icon of resistance against this trend. In the realm of the natural sciences, as John Dupré explains, "more recently it has been increasingly widely recognized that in practice no such reductive programme was remotely feasible" (2001: 6). He devotes a chapter of his book *Human Nature and the Limits of Science* to economics and argues against "its typical commitment to explanations of social-level economic phenomena by appeal to individual behavior" (2001: 131), which, he adds, shares the reductionist basic commitment to monocausality. This, Dupré states, is "too simplistic a view of the relation between the individual and the social" (2001: 153).

Only a few authors have undertaken a philosophical examination of the relations between micro- and macroeconomic entities—some of them will be mentioned in this chapter. The reductionist program denounced by Dupré has been unconsciously accepted by standard economics almost

R. F. Crespo, *The Nature of the Economy*,
https://doi.org/10.1007/978-3-031-02453-5_8

without any discussion. This chapter aims to make use of the few existing philosophical analyses about the microfoundations of the macroeconomics program and try to assess it by developing an ontology of macroeconomic entities. As Michiru Nagatsu states, "metaphysics is an indispensable part of scientific practice that provides scientists with worldviews and directions in research" (2010: 198). We need metaphysics to ascertain the nature of the subject matters of sciences to define their appropriate methods, scope, and limits.

The first section will introduce the definitions and differences between micro- and macroeconomics. The second section will present the microfoundations program and its core tool, the "Representative Agent." The third section will tackle the ontological nature of macroeconomic entities, considering the notions of supervenience and emergence to describe John Maynard Keynes's position before applying Aristotle's theory of wholes to the chapter's topic. Finally, a short conclusion will wrap up the previous arguments. We should bear in mind the distinction made in the previous chapter between methodological and ontological individualism and collectivism. The microfoundation program and the representative agent are methodological strategies or instruments that tacitly assume ontological conceptions. However, they can be combined in different ways: ontological individualists are mostly methodological individualists but can also refer to social interactions of individuals in their explanations. Ontological collectivists give methodological collectivist explanations but may also resort to individualist explanations.

Micro- and Macroeconomics

In his book *Causality in Macroeconomics*, Kevin Hoover starts the first chapter by defining micro- and macroeconomics. He states, "'Macroeconomics' is sometimes thought to be the economics of broad aggregates, and 'microeconomics' the economics of individual economic actions" (2001a: 108). In these definitions, while the meaning of microeconomics seems satisfactory, the meaning of macroeconomics proves rather vague—"is sometimes thought to be...." Hoover later tries to clarify this meaning by exemplifying the topics macroeconomics deals with: "Macroeconomics is thus the area of economics that studies GDP, unemployment, interest rates, the flow of financial resources, exchange rates, and so forth" (2001a: 109). Athol Fitzgibbons, in his book *The Nature of Macroeconomics* (2000), also describes it in a similar fashion: "macroeconomics concerns inflation, unemployment, money, international trade and economic growth," while adding its purpose: "the direct purpose of macroeconomic theory is to guide general policies that affect the immediate happiness and well-being of hundreds of millions of people" (2000: xi). I think this description constitutes a good contribution, since defining purpose helps circumscribe a subject.

The Oxford *Dictionary of Economics* states that microeconomics studies "The micro aspects of economics, concerning the decision-making of individuals. Microeconomics analyses the choices of consumers (who can be individuals or households) and firms in a variety of market situations. Its aim is to explore how choices should be made, and to provide an explanation of choices that are made" (Hashimzade et al. 2017).

The same dictionary states that macroeconomics studies:

> The macro aspects of economics, concerning the determination of aggregate quantities in the economy. Macroeconomics considers what determines total employment and production, consumption, investment in raising productive capacity, and how much a country imports and exports. It also asks what causes booms and slumps in the short run, and what determines the long-term growth rate of the economy, the general level of prices, and the rate of inflation. Macroeconomics considers how these matters can and should be influenced by government through monetary and fiscal policies. It is contrasted with microeconomics, which is concerned with disaggregated quantities, such as the incentives operating on individuals and firms in the economy, the organization of production, and the distribution of incomes. The emphasis on building macroeconomic models on the basis of microeconomic foundations has led to an amalgamation of the two in many instances.

I find that, while the definition of microeconomics sets forth a specific subject matter, the definition of macroeconomics has been reduced to a description of the topics it encompasses. However, this last quotation seems to establish the difference between micro- and macroeconomics by pointing to the respectively individual and aggregated nature of their subject matters. Microeconomics deals with economic decisions and actions of individual entities: individual persons, households, firms. It implicitly recognizes that, to some extent, "social collectives" as households and firms make decisions and perform actions. It treats them as if they were "moral persons" that decide and act. That is, microeconomics deals with the individual's (in an ample sense) economic decisions and actions. It seems that it does not consider government decisions.

Concerning the topics addressed by macroeconomics, I question their allegedly aggregated condition. Let us call them macroeconomic "entities" (I use the term "entity" in a general way that includes things, properties—all that exists). This aggregated condition seems applicable to some of them, but not all of them—for example, the exchange or interest rates. That is, some macroeconomic "entities" derive from the aggregation of individual data—for example, the GDP, unemployment rates, or the balance of payments. Yet others do not stem from an aggregation of units. In sum, we cannot define macroeconomics by the aggregated character of its entities (as it is often done). However, this does not mean that these entities do not emerge from human decisions or actions. Hence, we need to look for more adequate definitions for macroeconomics.

The U.S. Federal Reserve defines macroeconomics as "the study of whole economies—the part of economics concerned with large-scale or general economic factors and how they interact in economies [...] The Fed cares about macroeconomics because its goals are determined and defined in macroeconomic concepts."[1] I find this wholeness-based definition more comprehensive than the aggregate-based definition. The classical textbook on macroeconomics by Rudiger Dornbusch and Stanley Fisher also states that macroeconomics deals with "the behavior of the economy as a whole," with "the essentials," disregarding the details of the behavior of individual economic units (2018: 2). In sum, I agree that microeconomics deals with individual economic decisions and actions, including individual persons, families, and firms. As for macroeconomics, I suggest that it deals with the "economy as a whole," the entire economy, which is not the same as stating that it addresses the aggregate of economic individual actions or events. I will argue that, tough related, both subject matters—the individual economic decisions and actions as well as the whole economy—are different and irreducible.

[1] https://www.federalreserve.gov/faqs/what-is-macroeconomics.htm. See also Maarten Janssen (1993: 2–6), for different definitions of micro and macroeconomics. Janssen also mentions (1993: 5) an interesting distinction made by E. Roy Weintraub, who assigns the assumption of coherence to microeconomic outcomes and system-incoherence to macroeconomics: "the appropriate cut is between models of coordination success and models of coordination failure" (1979: 74–75).

The next logically subsequent questions are: What is "the economy as a whole"? What is its relationship with individual economic decisions and actions? The answers to these questions involve the ontological conditions of macro entities. However, before delving into this complex issue, I will describe the microfoundations of the macroeconomics project, because it calls for an analysis of the ontological nature of macro entities. That is, I am starting with a matter of method, and then I will address the metaphysic nature of macro entities.

Microfoundations and the Representative Agent

Economists usually believe that they have designed a satisfactory microeconomic theory. Yet, the new behavioral and experimental economics have discovered that this theory is not satisfactory enough. However, economists still think that they can improve or change the theory for an adequate result.

Meanwhile, despite numerous existing theories—Monetarist, Keynesian, Austrian, Neo-Classical, Neo-Keynesian, and other theories—macroeconomics still lacks a stable, uniform, well-established theory. Additionally, the Sonnenschein-Mantel-Debreu (SMD) theorem proves that it is not possible to achieve a unique and stable equilibrium when aggregating individual behaviors. As S. Abu Turab Rizvi (2006: 230) explains, "SMD theory means that assumptions guaranteeing good behavior at the microeconomic level do not carry over to the aggregate level or to qualitative features of the equilibrium."

At the same time, the individualist bias in the economic profession (different from the *ethos* of the sociological profession) makes standard economists "naturally" inclined to believe that social entities or events can be reduced to individual entities or events. This notion, added to the above-mentioned macroeconomic issues, "naturally" drives economists to look for the microfoundations of macroeconomics. It is a powerful program. As Robert Lucas famously put it, "the term 'macroeconomic' will simply disappear from use, and the modifier 'micro' will become superfluous" (1987: 117). This is the dominant view among economists today.[2] At the other end of the spectrum, Don Ross (2014) seems to practically reduce economics to macroeconomics.[3]

As Alan Kirman states (1992: 119), the easiest way to reduce macroeconomics to microeconomics is to consider the whole economy as an individual that behaves as individuals do (see also Janssen 2006: 4, 6). This is the representative agent in macroeconomic models.[4] Before introducing Kirman's criticism to this construct, I will mention some "previous" issues raised around it.

Kevin Hoover and Wade Hands draw attention to a first problem: the technical difficulties plaguing "aggregation" (Hands 2017: 1686; Hoover 2001c). There is a way of circumventing this issue: "to skip individual agents altogether and simply assume a single representative agent is the sole decision-maker in the model" (Hands 2017: 1686). A second problem revolves around the constraints that restrict the traits of a representative agent (Jackson and Yariv 2019: "The existence of a representative agent imposes strong restrictions on individual utility functions"; Jarrow & Larsson 2017; An et al. 2009).

Third, there is the issue of the notion of representative agent used: the conventional economic utility-maximizing agent. As already mentioned, lab and natural experiments, as well as behavioral economics have challenged this concept. As Hands states, "The empirical and theoretical adequacy of individual choice theory has increasingly been a topic of debate." Thus, he continues, "It seems ironic that at a time when utility-maximization is being questioned in its original domain of individual choice, it would be extended to predict and/or explain the behavior of entire markets and/or entire economies" (Hands 2017: 1687), and he asks, "Why extend the utility-maximizing agent to markets and whole economies at a time when the profession is so conflicted regarding the proper role of utility-maximization in the study of individual behavior?" (Hands 2017: 1693). Or is the position in favor of the representative agent assuming the classical liberal belief that irrational individual actions may lead to collective rational actions? (à la Mandeville).

[2] See, for example, the entry "Macroeconomics: Relations with Microeconomics" in the New Palgrave Dictionary of Economics (Holwit 2016).

[3] See the reviews by John B. Davis (2014) and Michiru Nagatsu (2015).

[4] Wade Hands (2017: 1687, footnote 4) shows the increasing presence of the expression in economic journals via a JStor search.

A fourth problem hinges on the fact that the representative agent model fails to take advantage of rich heterogeneity. For Kenneth Arrow, "the homogeneity assumption seems to me to be especially dangerous. It denies the fundamental assumption of the economy, that it is built on gains from trading arising from individual differences" (1986: 390), while, in a letter to Samuelson, Frank Hahn asserts, "Rather it is the use of the 'representative agent'—a disastrous concept. Certainly no way to start an analysis of 'coordination' failures... It is not General Equilibrium analysis" (quoted by Hands 2017: 1692). In his Nobel Prize Lecture (December 8, 2000), James Heckman also noted:

> The most important discovery [from microeconometric investigations] was the evidence on the pervasiveness of heterogeneity and diversity in economic life. When a full analysis of heterogeneity in responses was made, a variety of candidate averages emerged to describe the "average" person, and the long standing edifice of the representative consumer was shown to lack empirical support. (2000: 256)

Sheila Dow (2021) also remarks the macroeconomics' struggle between theoretical and empirical coherence. She states in philosophical terms:

> For modern mainstream macroeconomics, the implicit ontology takes agents to be narrowly rational, optimizing, and atomistic beings, facilitating formal deductivist expression. This justifies the strict form of microfoundations as separable and unidirectionally causal. Yet the confrontation of the resulting (closed-system) macroeconomic theory with the (open-system) context of its application opens mainstream macroeconomics to the charge of incoherence; (classical) logical consistency conflicts with philosophical consistency. (2021: 13–14)

Returning to Kirman (1992), he notes and demonstrates several flaws in reducing macroeconomics to microeconomics with the "representative agent" construct. First, there is no direct relationship between individual and collective behavior; second, the reaction of the representative to some change in a parameter may differ from the reaction of the aggregate of agents represented; third, preferences may also be different; finally, "The sum of the behavior of simple economically plausible individuals may generate complicated dynamics, whereas constructing one individual whose behavior has these dynamics may lead to that individual having very unnatural characteristics" (1992: 118). Kirman concludes that "the assumption of a representative individual is far from innocent; it is the fiction by which macroeconomists can justify equilibrium analysis and provide pseudo-microfoundations" (1992: 125). He further points out that "the representative agent approach is fatally flawed because it attempts to impose order on the economy through the concept of an omniscient individual. In reality, individuals operate in very small subsets of the economy and interact with those with whom they have dealings" (1992: 132). He believes that considering the differences between these subsets may provide better results.

This fiction is inadequate, not only due to the aforementioned issues raised by Kirman but also because of the defective character of the theory of the individual adopted, as noted. This might be why Kirman speaks of "pseudo-microfoundations." When he calls for non-pseudo foundations, what he wants is actual, diverse individuals: in the real world, individuals are inescapably different—though not simply physically.

The rational maximizer adopted as representative agent seems to have "supernatural powers," as Diana Patrascu (2013: 42) puts it.[5] Nonetheless, even assuming it is possible to improve the individual economic agent's characterization, the behavior of this renewed representative agent will not match collective behavior.[6] People's interactions add reactions and changes that cannot be considered individually. In addition, not all macroeconomic entities are linked to individual agents' decisions and actions.

The heterogeneity of agents and their dynamic coevolution and interaction are considered by the new "agent-based modeling" methods. The literature on them is immense and expanding.[7] It would be premature to

[5] Kenneth Boulding also uses a religious expression—"the Immaculate Conception of the Indifference Curve" (1969: 2)—to highlight the standard economic agent's lack of realism.

[6] Kirman sums up (1992: 134), "Given the arguments presented here—that well-behaved individuals need not produce a well-behaved representative agent; that the reaction of a representative agent to change need not reflect how the individuals of the economy would respond to change; that the preferences of a representative agent over choices may be diametrically opposed to those of society as a whole—it is clear that the representative agent should have no future.".

[7] See for example http://www2.econ.iastate.edu/tesfatsi/amulmark.htm, for an updated literature review on agent-modelling, as well as the *Journal of Evolutionary Economics* 29/1 (2019) issue, starting with its introduction by Giovanni Dosi and Andrea Roventini.

offer a definitive judgment about the possibility of these methods to overcome the problems of the already old representative agent theory. While recognizing that these models constitute a significant advantage over representative agent models, Brian Epstein (2011) shows their limitations and concludes that we should give up the assumption that agent-based models can provide the microfoundations for macroproperties (2011: 115, and *passim*): these properties are not fully determined (or exhausted by or entailed by) individualistic properties (2011: 119). He concludes that we have to explicitly include macro entities in these models (2011: 132–133).[8]

At this point, the time has come to tackle the ontological matter. By ascertaining the nature of macro entities, it will shed light on the method to analyze them.

The Ontological Nature of Macroeconomic Entities

I will first explain "supervenience" and then describe the "emergence" theory. Finally, I will introduce Keynes' and Aristotle's positions, which are connected to the latter.

Supervenience

What is the ontological nature of the GDP; the balance of payments, and the inflation, interest, or exchange rates? Do they really exist, or are they only human creations? As Kevin Hoover asks, "*Do macroeconomic aggregates and causal relations among them exist?* More precisely [...], do macroeconomic aggregates exist *externally* (i.e. independently of any individual mind) and *objectively* (i.e., unconstituted by the representations of macroeconomic theory)" (2001a: 109; italics in the original).

Hoover maintains that macroeconomic entities really exist, but they are linked to microeconomic entities. He resorts to the concept of supervenience to defend the reality of macro entities without reducing them to micro entities. "Supervenience" is a philosophical concept that has been widely used in different branches of philosophy, especially in metaphysics, ethics, and the philosophy of mind. "A set of properties *A* supervenes

[8] See also John Davis (2011: 165).

upon another set *B* just in case no two things can differ with respect to *A*-properties without also differing with respect to their *B*-properties" (Brian McLaughlin and Karen Bennett 2011: 1). For example, while mental properties emerge from physical properties, they are distinct. Donald Davidson states:

> [M]ental characteristics are in some sense dependent, or supervenient, on physical characteristics. Such supervenience might be taken to mean that there cannot be two events alike in all physical respects but differing in some mental respect, or that an object cannot alter in some mental respect without altering in some physical respect. (1980, p. 214)

Regarding economics, Hoover notes:

> Macroeconomic aggregates *supervene* upon microeconomic reality. What this means is that, even though macroeconomics cannot be reduced to microeconomics, if two parallel worlds possessed exactly the same configuration of microeconomic or individual economic elements, they would also possess exactly the same configuration of macroeconomic elements. It is not the case, however, that the same configuration of macroeconomic elements implies the same configuration of microeconomic elements. (2001a: 120)

That is, "the mapping is not one to one" (2001c: 73). For Hoover, macroeconomic elements exist, but they would not exist without a substrate of microeconomic elements (2001a: 124). He offers two arguments for the real existence of macroeconomic elements: first, they can be manipulated, and second, they are captured by models that effectively explain empirical phenomena—"to the degree that such theories are empirically successful, the best account of these macroeconomic entities is that they are real" (2001b: 244). Other philosophers of economics also apply supervenience to the micro-macro case. For example, Wade Hands briefly describes and adheres to this position (2001: 171).

Some authors object to this position. Harold Kincaid (1996) makes an argument against microfoundations in the context of his "empirical case against individualism." He lists the following obstacles to reduction: (1) the possibility of multiple realizations—social phenomena that "can be brought about in diverse ways by individual behavior in different situations or at different times" (1996: 154); (2) the usual need to presuppose social elements when describing individuals; (3) the existence of many relations between individual and social elements that impede identifying specific explanations, and (4) a "context sensitivity" that allows for different characterizations of individual conducts depending on social contexts. Kincaid provides examples of rational choice and game theories, and he also applies these objections to the microfoundation program (1996: 250–257).

Kincaid is reluctant to engage in metaphysical discussions, but he cannot help it. Social theories' best explanations refer to social entities: thus, "we are committed to their existence. Moreover, we have argued that such entities can stand in social relations, both to other entities and to individuals. So they are 'causally real' as well" (1996: 188). This author rejects the supervenience thesis version that matches the "exhaustion" thesis (everything that exists is individual or an aggregate of individuals). Also, he adds, "aside from individuals, society might be composed of material goods as well" (1996: 188).

Brian Epstein's (2014) argument against the supervenience of macroeconomic properties on microeconomic ones refers to the notion stated above. For him, macroeconomic properties "depend, in part, on factors that neither are properties of individuals, even including their 'wide' attitudes, nor have close causal relations with individuals" (2014: 10–11).[9]

Julian Reiss (2004) also argues against supervenience because of its reductionism. In fact, Hoover (2009: 390) clarifies that he uses supervenience "in an anti-reductionist manner," and he incorporates John Searle's (1995) account of social facts as an adequate framework "to provide an account that both assigns an independent ontological status to microeconomic individuals and to macroeconomic aggregates and provides an intelligible account of the connection of the intentional states of the individuals and the behavior of aggregates" (2009: 405).

Emergence

The concept of "emergence" arises as a "weaker" notion than supervenience (in the sense that macro entities are less dependent on micro entities) to explain the relation between micro and macro elements.

[9] For a critique on supervenience from another point of view, see Julie Zahle (2021) and Petri Ylikoski (2014). Supervenience is not a unified theory. As Ylikoski affirms, "there are dozens of forms" of it (2014: 119). See also Chorafakis (2020: 241–242).

Timothy O'Connor and Hong Yu Wong (2020) have described the non-reducible character of emergentism: "emergent entities (properties or substances) 'arise' out of more fundamental entities and yet are 'novel' or 'irreducible' with respect to them. (For example, it is sometimes said that consciousness is an emergent property of the brain.)" The roots of emergentism lie with British philosophers. An epistemological or weak version of emergentism derives from Samuel Alexander, while its ontological or strong version can be traced back to John Stuart Mill and C.D. Broad.[10]

Ontological emergentism holds that emergent properties are *causal interactions* that are additional to the fundamental entities from where they emerge. For ontological emergentists, emergent laws have not only same-level effects but also effects in lower levels, i.e., a "downward causation."

According to Maarten Janssen (2006: 2),

> An important early example of emergence is the analysis of Schelling (1978) on segregation. He shows that segregation in neighborhoods may be an emergent property at the micro level that can be viewed as an unintended consequence of the individual decisions concerning where to live.

Janssen himself recognizes the existence of these emergent properties (2006: 8). Indeed, Schelling asserts that the situations.

> in which people's behavior or people's choices depend on the behavior or the choices of other people, are the ones that usually don't permit any simple summation or extrapolation to the aggregates. To make that connection we usually have to look at the *system of interaction* between individuals and their environment, that is, between individuals and other individuals or between individuals and the collectivity. (1978: 14, my italics)

Schelling's intention is not ontological, but he is implicitly providing an ontological clue when he stresses the need to look at "the system of interaction" between individuals, the environment, and collectivity. Here

[10] This section does not intend to discuss the many versions of emergence considered in the specific literature. For a brief and useful summary, see Chorafakis (2020: 242–245).

there is a "plus" that we will fail to capture if we only look at individual behaviors.

As explained in the previous chapter, Dave Elder-Vass defends a "relational theory of emergence" and the ascription of causal powers to social entities (2010: 14ff.). For him, there are wholes—structures—that consist of "the set of relations between the parts that only exist when this structure is present" (2014: 45). These relations cause interactions that produce causal effects. He states that "entities at each level can have emergent causal powers: powers to affect the world that would not be possessed by the parts if they were not organized into entities of this nature" (2014: 45). Applied to social collectives, Elder-Vass claims that "As a result of its members being committed to interact in the ways specified in their roles, the organization has the capacity to have a causal impact on the world that its members would not have if they were not parts of the organization concerned" (2014: 49). He (2014: 2017) argues that social entities are often composed of both human agents and non-human material objects, and that both may make essential contributions to their causal influence. This author has recently (2016) applied his theory to digital economy, explaining the success of some web sites as the result of the interaction between users and technological devices with emergent causal powers that none of them would possess outside such combination.

As noted in Chapter 3, Tony Lawson (2016: 963–96) also maintains that social reality largely consists of emergent totalities, and that "ontological reduction of any emergent totality to the pre-existing elements alone, considered apart from their being relationally organized, is proscribed [...] The structure makes a difference" (2016: 965).

Raimo Tuomela applies the idea of emergence to social collectives, and he connects it to the purposes of macroeconomics (2013: 46, 223). Tuomela thinks that the social realm cannot be explained by supervenience (which implies parallel properties and changes in the micro- and macro-levels) but by emergence (2013: 16, 21, 53, 91). For him, a central factor for the emergence of a social group is to have a reason for action (2013: 11, 34, 38ff.) or an *ethos* (2013: 16, 26, 30, 34). This reason or *ethos* triggers the need for functions and actions (2013: 1, 2, 8, 16, 44).

James Hartley also supports the notion of emergent macroeconomic entities (1997: 188ff). His claim draws from James Coleman, whom he quotes (1990: 28):

> the only action takes place at the level of individual actors, and the "system level" exists solely as emergent properties characterizing the system of action as a whole. It is only in this sense that there is behavior of the system. Nevertheless, system-level properties will result, so propositions may be generated at the level of the system.

That is, for Coleman, actions are individual, but interactions add something else. He states: "the social system consists of the combination of these actions of independent individuals… [T]he fiction is just that—for individuals do not act independently, goals are not independently arrived at, and interests are not wholly selfish" (1990: 300–301).

Thus, institutions are crucial because they embody common ends and regulate the ways in which they can be achieved. As Searle affirms, "An institution is any collectively accepted system of rules (procedures, practices) that enable us to create institutional facts" (2005: 21).[11] The economic system is full of institutions: money, banks, firms, corporations, government regulations…[12] Interactions guided by institutions add something that cannot be accomplished by isolated individuals. Hartley argues:

> a firm is a system that organizes the actions of multiple individuals so that the resulting behavior is very different from what would be the behavior of a single individual. So, when we turn to the matter of macroeconomics, it does not seem such an oddity that we might think that the whole (the macroeconomy) is more than the sum of its parts (people and firms). (1997: 186)

Margaret Schabas supports this notion with the views expressed by Friedrich List, John Stuart Mill, even Friedrich A. von Hayek and John Searle, and she concludes that "insofar as 'the economy' is an emergent property of our world, it resists reduction down to individual agency" (2009: 16).

[11] I will not delve into the concept of institution here.

[12] Searle's views on the primacy of collective intentionality characterize him as a holist (see the previous chapter), but he is an ontological individualist.

Keynes on the Economy

The idea that the whole is greater than the sum of its parts is traditionally held by social "organicists," with its most extreme proponents viewing the social whole as a substance. However, other sociologists, such as Coleman, hold more moderate views that do not support the notion of the social whole as a substance. From his early intellectual days, John Maynard Keynes, influenced by Franz Brentano and George E. Moore, adopted an organic view of the good and of psychical phenomena which probably impacted on his economic ideas. Both Brentano and Moore maintained a moderate organicist position. Moore and Keynes read Brentano's *The Origin of the Knowledge of Right and Wrong*, where the author approvingly states: "Aristotle called attention to the fact that the total sum of what is good is always better than the particular parts that make it up" ([1889] 2009: 17). Brentano notes that, for Aristotle, the whole is not a substantial whole but an ordered, consequently accidental whole ([1889] 2009: 25). Explicitly following Brentano, Moore states ([1902] 1966: xi):

> I shall, where it seems convenient, take the liberty to use the term 'organic' with a special sense. I shall use it to denote the fact that a whole has an intrinsic value different in amount from the sum of the values of its parts. I shall use it to denote this and only this. The term will not imply any causal relation whatever between the parts of the whole in question. And it will not imply either, that the parts are inconceivable except as parts of that whole, or that, when they form parts of such a whole, they have a value different from that which they would have if they did not. ([1902] 1966: 35–36)

Keynes recalled his adherence to this position in his 1938 essay "My Early Beliefs," where he calls this view "the principle of organic unity" (1972a: 436–437), an expression borrowed from Moore ([1903] 1966: xi). He refers to it in his *Treatise of Probability* (1921), where he considers the possibility of applying it to the physical realm (1921: 249) and, more resolutely, to the human realm (1921: 310, 355, 346). Again, this view surfaces in his biographical sketch of Frank Ysidro Edgeworth (written in 1926):

> The atomic hypothesis which has worked so splendidly in physics breaks down in psychics. We are faced at every turn with the problems of organic

> unity, of discreteness, of discontinuity—the whole is not equal to the sum of the parts, comparisons of quantity fail us, small changes produce large effects, the assumptions of a uniform and homogeneous continuum are not satisfied. (1972b: 262)

In his *The End of Laissez-Faire* (1927), Keynes applies the organic view to production and consumption processes:

> The beauty and the simplicity of such a theory [the *Laissez-Faire* theory] are so great that it is easy to forget that it follows not from the actual facts, but from an incomplete hypothesis introduced for the sake of simplicity. Apart from other objections to be mentioned later, the conclusion that individuals acting independently for their own advantage will produce the greatest aggregate of wealth, depends on a variety of unreal assumptions to the effect that the processes of production and consumption are in no way organic, that there exists a sufficient foreknowledge of conditions and requirements, and that there are adequate opportunities of obtaining this foreknowledge. (1927: 32–33)

These ideas are also present in his *General Theory*. Keynes often speaks of complexity in the cases of aggregate entities—for instance, "the community's output of goods and services is a non-homogeneous complex which cannot be measured, strictly speaking, except in certain special cases, as for example when all the items of one output are included in the same proportions in another output" ([1936] 1973: 38). Originally trained in mathematics, his view of its limited usefulness is also associated with this complexity: "Too large a proportion of recent 'mathematical' economics are merely concoctions, as imprecise as the initial assumptions they rest on, which allow the author to lose sight of the complexities and interdependencies of the real world in a maze of pretentious and unhelpful symbols" ([1936] 1973: 298).

Therein lies the root of Keynes' emphasis on dealing with the "wholes." In a letter to Roy Harrod (Keynes to R. F. Harrod, 30 August 1936, volume xiii), he states, "To me the most extraordinary thing, regarded historically, is the complete disappearance of the theory of demand and supply for output as a whole, i.e. the theory of employment, after it had been for a quarter of a century the most discussed thing in economics." Keynes also warns about this in the original Preface of the *General Theory* (xvi), and in the Preface to the German edition, anticipating that he would be "offering a theory of employment and output as a whole" (xix; and in the Preface to the Japanese and French editions, xx and xxii). In the French edition, he also adds, "I argue that important mistakes have been made through extending to the system as a whole conclusions which have been correctly arrived at in respect of a part of it taken in isolation" (xxii). The expression "the community [or labor, industry, employment, output, effective demand] as a whole" is constantly repeated throughout the book. In sum, Keynes, probably influenced by his philosophical studies, believes that the wholes should be considered separately from their parts, given that the behavior of a whole differs from what we can infer from individual behavior. His view goes beyond the scope of supervenience and is closer to the idea of emergence.

Aristotle on Wholes and Parts

Brentano refers to Aristotle's vision of ordered wholes, and I believe this position may provide an ontological account of macroeconomic entities associated with Keynes' vision.

In his entry "Emergent Properties" at the *Stanford Encyclopedia of Philosophy,* Timothy O'Connor and Wong (2020) states that "Although debates concerning the reality or precise nature of emergence are largely driven by contemporary scientific theorizing, the basic notion has quite a long history stretching back at least to Aristotle (384–322 BC)" (2020: 1). Specifically, Aristotelian scholars considering him as the first "emergentist" quote the passage from *Metaphysics* mentioned in the previous chapter: "In the case of all things which have several parts and in which the totality is not, as it were, a mere heap, but the whole is something besides the parts, there is a cause" (*Metaphysics* VIII, 6, 1045a 9–10). O'Connor adds that "Aristotle's form/matter compound conception of material substances is consonant with the standard emergentist stance between substance dualism and reductionism." Form and matter are not substances but co-principles of substances, which are something in addition to being themselves.

Let me briefly recall that, as explained in the previous chapter, for Aristotle, unities of things or wholes may be substantial, merely accidental, or unities of order. Between the strongest—the substantial—and the accidental wholes, there are unities of order when the parts preserve their substantiality and ends, but the order also allows each part to contribute to a common end. Also, the expression "unity of order" applied to society has been drawn from Thomas Aquinas. In his *Commentary to the Nicomachean Ethics,* Aquinas affirms that Aristotle's conception of society as a whole has to be conceived as wholes that possess not an "absolute unity," but a "unity of order" (*In Ethic*, I, 1, 5; *Contra Gentes*, IV, 35).[13] In his *Summa Theologiae,* he states that "The community is not a *supposit*, an *unum* per se. Thus, its proper effect is orderly relations" (III, 4, a.1, ad 4). In other parts of his *opus,* Aquinas applies this expression to the order of the world in reference to Aristotle's *Metaphysics* XII, 10. There, Aristotle writes that all things of the universe are ordered together somewhat, but not all alike; they are connected: "for all are ordered together to one end" (1075a 21). Aquinas uses the examples of an army and of a household and their members. The root and nature of the unity of the whole is the end and the consequent order, but it is not a substantial whole.

As also noted in the previous chapter, specifically referring to the nature of society and contrary to Plato's monistic view, Aristotle denies that the *polis* is a substantial unity: "it would be the destruction of the polis," he sentences (*Politics* II, 2, 1261a 24–25). He also denies the sophists' concept of the *polis* as a merely accidental unity (*Politics* III, 3). For him, the *polis* is a whole composed of substantial parts that interact in an orderly fashion. These interactions would not necessarily be performed by individuals out of society. There is something ontological in that the behavior of the whole differs from a generalization of individual behavior.

We can apply this idea of ordered wholes to the economic system. The order may succeed or fail; coordination is not always ensured. However, there is a set of institutions that promotes it through their regulations. An economic system emerges from the micro-level with its own laws, actions, and tendencies that have repercussions at the micro-level. My colleague Daniel Heymann suggested that I should read Phil Anderson's article published in *Science* with a title that says it all: "More is Different" (1972). Though Anderson refers to examples taken from physics (thus probably dealing with substantial wholes), Heymann applies his idea to the economy. Anderson affirms, "at each level of complexity entirely

[13] "Sciendum est autem, quod hoc totum, quod est civilis multitudo, vel domestica familia, habet *solam unitatem ordinis*, secundum quam non est aliquid simpliciter unum." He uses the example of the army and its members, and he adds: "Est autem aliquid totum, quod habet unitatem non solum ordine, sed compositione, aut coligatione, vel etiam continuitate, secundum quam unitatem est aliquid unum simpliciter, et ideo nulla est operatio partis quae non sit totius" (*In Eth.*, Lectio I, Caput 1, 5). In this way he distinguishes the unity of order from other kinds of unities or wholes.

new properties appears [...] At each stage entirely new laws, concepts and generalization are necessary" (1972: 393).

Conclusion

Having presented different theories to explain why some kinds of wholes are not the mere result of aggregating parts, it is now time to sum up these ontological positions as applied to macroeconomics. The substantial entities implied in the macroeconomy are individuals and other substantial things that generate or influence macroeconomic entities. These entities create, in Coleman's words, a "system level" that exists as a net of properties of individual and other substantial relevant things and relations among them. It is in this sense that the system has a behavior: the economy, the GDP, the demand for money or employment may or may not grow. The macroeconomy is not a substantial reality, a "big thing," but a system of interactions guided by institutions that has a "life" of its own. It is an accidental entity that ultimately inheres in the micro entities but cannot be reduced to them.

The emergence theory often speaks about levels—a lower and a higher level that emerges from the former. I think that it is not necessary to consider different "hierarchical" levels[14]: they are just different ontological realities. Though consisting of and inhering in micro entities, the economic system—people, interactions among them, institutions, conventions, beliefs, other material stuffs—can be considered as a different entity.

Which of these realities has an ontological priority? Here, we can apply the conclusions about society in general described in the previous chapter. Obviously, persons who are substances, and their actions have a greater

[14] Zahle (2021) and Ylikoski (2014) are also against considering levels, but they criticize the emergence theory as well. I think that one thing does not imply the other. For Searle, there are "ground-floor" facts about the economy—buying, selling, and other economic activities and attitudes of people. Searle states that "these have certain macro consequences such as, for example, the trade cycle. But the systematic fallouts are macro facts that are all constituted by the ground-floor or lower-level institutional facts [...] (2010: 22). For him, ground-floor facts are usually intentional, while macroeconomic facts are typically "intentionality-independent" (2010: 117). He uses the example of the 1929 Great Depression. We could dissect it and analyze its components and their interactions, but the reconstruction will not be the Great Depression. He states, "In general, one can say that ground-floor institutional facts are studied by microeconomics, systematic fallouts by macroeconomics" (2010: 22).

ontological "density" that the economic system, which is a dimension of the social system. However, as also explained in the previous chapter, people's final end is to live in society, and, given that Aristotle regards the final as the cause of causes, society has thus an ontological priority over its components. Can we apply this conclusion also to the economic system? I think that the case is different, because it does not seem that microeconomic actions are naturally oriented to the economic system as their final end.

What is the consequence of this ontological characterization for the economic science? Microeconomics deals with individual decisions and actions, which might be prompted by maximizing utility intentions, psychological bias, or drivers that belong to the fields of behavioral economics and neuroeconomics. The macroeconomic approach is very different: it includes these individual conducts, even though these microeconomic reasons to act belong in fact to a theoretical fictional model. They can be considered and will provide a heuristic orientation about individual behavior. However, the fact that people live in a society with common ends fostered by institutional arrangements makes them behave and interact in different ways than they would if they were alone. As Reiss notes, "clearly, macro entities causally influence micro entities" (2004: 232; a point also made by Kincaid 1996: 253). Hoover states that "macroeconomic concepts are, in fact, constitutive of parts of microeconomic reality" (2009: 391), because individual agents employ macroeconomic concepts in their decisions. Additionally, as already noted, Epstein (2011: 132–133) introduces macro entities in agent-based models. More emphatically, Don Ross argues that microeconomics' foundations lie in macroeconomics (cf. 2014: 89).

In addition, there are not only individual behaviors as well as institutional rules and policies but also natural or human exogenous conditions—natural resources or culture—and events—a pandemic or a war—that condition or produce economic consequences that cannot be explained by individual economic decisions. The macroeconomy is a whole, comprising individuals interacting in an institutional context plus other conditions, facts, or events that are not managed by individuals. There is an ontological difference between an individual acting alone and the performance of the whole economy. They are different kinds of stuff. This is the reason why I agree with John E. King. He suggests:

> microeconomics and macroeconomics are related to each other horizontally, not vertically. The phenomena that they study are rather obviously closely related, and their practitioners most certainly can (and should) cooperate with each other and learn from each other. But neither should be seen as the 'foundation' of the other [...] Microeconomics and macroeconomics should be thought of as two separate buildings, equal in height, adjacent to each other and connected by footbridges. Both rest on a single, solid and extensive set of social and philosophical foundations. (2012: 10 and 26)

In this regard, King quotes (2012: 44) Kriesler: "Rather than any form of hierarchical relationship, the two theories lie side by side (so to speak), and both give information which the other cannot give, while the interrelation of the two yields further information not obtainable from either in isolation" (Kriesler 1989: 123).

This chapter starts with a quote from John Dupré on economics' simplistic view of the relation between the individual and the social. An ontological analysis of macroeconomic entities shows that they are more than individual decisions and actions and that, therefore, we cannot reduce the former to the latter. The macroeconomic stuff is a system that consists of interactions between individuals and their environment, featuring properties and behaviors that cannot be reduced to an aggregation of individual behaviors. This implies that micro- and macroeconomics as disciplines cannot be reduced to each other. Depending on the subject matter or the aim of the investigation, the study should focus on one or the other. Nonetheless, this does not entail a complete separation between them; in fact, they should seek to effectively interact without trespassing their respective boundaries established by their ontologically different subject matters.

References

An, S., Y. Chang and S.-B. Kim (2009). "Can a Representative-Agent Model Represent a Heterogeneous-Agent Economy?" *American Economic Journal: Macroeconomics*, American Economic Association 1/2: 29–54.

Anderson, P. W. (1972, August 4). "More Is Different", *Science* 177/4047: 393–396.

Aquinas. (1964). *In Decem Libros Ethicorum Aristotelis ad Nicomachum Expositio.* Torino: Marietti.

Aquinas. *Summa Theologica*. http://www.newadvent.org/summa/. Accessed 27 February 2021.

Aristotle. (1984). *The Complete Works of Aristotle*, edited by Jonathan Barnes. Princeton: Princeton University Press.

Arrow, K. J. (1986). "Rationality of Self and Others in an Economic System", *Journal of Business* 59: 385–399.

Boulding, K. (1969). "Economics as a Moral Science", *The American Economic Review* 59/1: 1–12.

Brentano, F. ([1889] 2009). *The Origin of the Knowledge of Right and Wrong*, translated by Roderick Chisholm and Elizabeth Schneewind. London: Routledge (*Vom Ursprung sittlicher Erkenntnis*. Leipzig: Duncker & Humblot, 1889).

Chorafakis, G. (2020). "Emergence Versus Neoclassical Reductions in Economics", *Journal of Economic Methodology* 27/3: 240–262.

Coleman, J. S. (1990). *Foundations of Social Theory*. Cambridge, MA: Belknap Press.

Davidson, D. (1980). "Mental Events". In: *Essays on Actions and Events*. Oxford: Clarendon Press, pp. 207–225.

Davis, J. B. (2011). *Individuals and Identity in Economics*. Cambridge: Cambridge University Press.

Davis, J. B. (2014). "Review of Don Ross's *Philosophy of Economics*", *Erasmus Journal for Philosophy and Economics* 7/2: 142–148.

Dornbusch, R. and S. Fisher (2018). *Macroeconomics*. New York: McGraw-Hill, 12th edition.

Dow, S. C. (2021). "The Pursuit of Coherence in Mainstream Macroeconomic Methodology" (December 14, 2021). Capitalism & Society, 15/1, Available at SSRN: https://ssrn.com/abstract=3985589. Accessed 25 January 2022.

Dupré, J. (2001). *Human Nature and the Limits of Science*. Oxford: Oxford University Press.

Epstein, B. (2011). "Agent-based Modelling and the Fallacies of Individualism". In: P. Humphreys and C. Imbert (eds.) *Models, Simulations, and Representations*. Routledge, pp. 115–144.

Epstein, B. (2014). "Why Macroeconomics Does Not Supervene on Microeconomics", *Journal of Economic Methodology* 21/1: 3–18.

Elder-Vass, D. (2010). *The Causal Power of Social Structures. Emergence, Structure and Agency*. Cambridge: Cambridge University Press.

Elder-Vass, D. (2014). "Social Entities and the Basis of Their Power". In: J. Zahle and F. Collin (eds.) *Rethinking the Individualism-Holism Debate. Essays in the Philosophy of Social Science*. Dordrecht: Springer, pp. 39–54.

Elder-Vass, D. (2016). *Profit and Gift in the Digital Economy*. Cambridge: Cambridge University Press.

Fitzgibbons, A. (2000). *The Nature of Macroeconomics*. Cheltenham: Edward Elgar.

Fodor, J. A. (1974). "Special Sciences (Or: The Disunity of Science as a Working Hypothesis)", *Synthese* 28/2: 97–115.

Hands, D. W. (2001). *Reflection without Rules*. Cambridge: Cambridge University Press.

Hands, D. W. (2017). "Conundrums of the Representative Agent", *Cambridge Journal of Economics* 41/6: 1685–1704.

Hartley, J. E. (1997). *The Representative Agent in Macroeconomics*. London: Routledge.

Hashimzade, N., G. Miles and J. Black (2017). *A Dictionary of Economics*. Oxford: Oxford University Press. https://www.oxfordreference.com/view/10.1093/acref/9780199237043.001.0001/acref-9780199237043. Accessed 30 September 2020.

Heckman, J. J. (2000). "Microdata, Heterogeneity and the Evaluation of Public Policy", Prize Lecture, December 8, 2000. https://www.nobelprize.org/uploads/2018/06/heckman-lecture.pdf. Accessed 30 September 2021.

Hoover, K. D. (2001a). *Causality in Macroeconomics*. Cambridge: Cambridge University Press.

Hoover, K. D. (2001b). "Is Macroeconomics for Real?". In: U. Mäki (ed.) *The Economic World View*. Cambridge: Cambridge University Press, pp. 225–245.

Hoover, K. D. (2001c). *The Methodology of Empirical Macroeconomics*. Cambridge: Cambridge University Press.

Hoover, K. D. (2009). "Microfoundations and the Ontology of Macroeconomics". In: H. Kincaid and D. Ross (eds.) *Philosophy of Economics*. Oxford: Oxford University Press.

Howitt, P. (2016). "Macroeconomics: Relations with Microeconomics". In: M. Vernengo, E. Pérez Caldentey and B. J. Rosser Jr (eds.) *The New Palgrave Dictionary of Economics,* Living Edition. https://link.springer.com/referenceworkentry/10.1057/978-1-349-95121-5_691-1. Accessed 20 September 2020.

Jackson, M. O. and L. Yariv (2019). "The Non-Existence of Representative Agents" (May 12, 2019), SSRN-id2684776.pdf.

Janssen, M. C. W. (1993). *Microfoundations. A Critical Inquiry*. London: Routledge.

Janssen, M. C. W. (2006). "Microfoundations", TI 2006–041/1 Tinbergen Institute Discussion Paper.

Jarrow, R. A. and M. Larsson (2017). "On Aggregation and Representative Agent Equilibria" (June 12, 2017). SSRN-id.

Keynes, J. M. (1921). *Treatise on Probability*. London: MacMillan.

Keynes, J. M. (1927). *The End of Laissez-Faire*. Published by Leonard & Virginia Woolf at the Hogarth Press, London, third impression.

Keynes, J. M. (1972a). "My Early Beliefs". In: *Essays in Biography, The Collected Writings of John Maynard Keynes*, Volume X, London: MacMillan, pp. 433–450.

Keynes, J. M. (1972b). "Francis Ysidro Edgeworth", *The Economic Journal*, March 1926. In: *Essays in Biography, The Collected Writings of John Maynard Keynes*, Volume X, London: MacMillan, pp. 251–266.

Keynes, J. M. ([1936] 1973). *The General Theory of Employment, Interest and Money*. In: *The Collected Writings of John Maynard Keynes*, Volume VII, London: MacMillan.

Kincaid, H. (1996). *Philosophical Foundations of the Social Sciences*. Cambridge: Cambridge University Press.

King, J. E. (2012). *The Microfoundation Delusion. Metaphor and Dogma in the History of Macroeconomics*. Cheltenham: E. Elgar.

Kirman, A. (1992). "Whom or What Does the Representative Individual Represent?", *Journal of Economic Perspectives* 6/2: 117–136.

Kriesler, P. (1989), "Microfoundations: A Kaleckian perspective". In: J. E. King (ed.) *An Alternative Macroeconomic Theory: The Kaleckian Model and Post-Keynesian Economics*. Boston, MA: Kluwer, pp. 55–72.

Lawson, T. (2016). "Social Positioning and the Nature of Money", *Cambridge Journal of Economics* 40/4: 961–996.

Lucas, R. (1987). *Models of Business Cycles*. Oxford: Blackwell.

McLaughlin, B. and K. Bennett (2011). "Supervenience". In: *The Stanford Encyclopedia of Philosophy*, E. N. Zalta (ed.) http://plato.stanford.edu/entries/supervenience/. Accessed 30 September 2020.

Moore, G. E. ([1903] 1966). *Principia Ethica*. Cambridge: Cambridge University Press.

Nagatsu, M. (2010). "Function and Mechanism. The Metaphysics of Neuroeconomics", *Journal of Economic Methodology* 17/2: pp. 197–205.

Nagatsu, M. (2015). "Book Review of Don Ross (2014) *Philosophy of Economics*, Palgrave", *Journal of Economic Methodology* 22/1: 123–128.

O'Connor, T. and H. Y. Wong (2020). "Emergent Properties". In: E. N. Zalta (ed.) *Stanford Encyclopedia of Philosophy*. http://plato.stanford.edu/entries/properties-emergent/.

Patrascu, D. (2013). "The Representative Economic Agent—An Epistemological Approach", *Annals of the "Constantin Brancusi", Economy Series* 2/2013: 42–46.

Reiss, J. (2004). "The Methodology of Empirical Macroeconomics by Kevin D. Hoover. Cambridge University Press 2001, xii + 186 pages", *Economics and Philosophy* 20/1: 226–233.

Rizvi, S. A. T. (2006). "The Sonnenschein-Mantel-Debreu Results after Thirty Years", *History of Political Economy* 38/suppl. 1: 228–245.

Ross, D. (2014). *Philosophy of Economics*. Houndmills: Palgrave-Macmillan.

Schabas, M. (2009). "Constructing 'the Economy'", *Philosophy of the Social Sciences* 39/1: 3–19.
Schelling, T. (1978). *Micromotives and Macrobehavior.* New York: Norton and Co.
Searle, J. R. (1995). *The Construction of Social Reality.* New York: Free Press.
Searle, J. R. (2005). "What Is an Institution?", *Journal of Institutional Economics* 1/1: 1–22.
Searle, J. R. (2010). *Making the Social World. The Structure of Human Civilization.* Oxford and New York: Oxford University Press.
Stoljar, D. (2015). "Physicalism", *The Stanford Encyclopedia of Philosophy*, E. N. Zalta (ed.) https://plato.stanford.edu/archives/win2017/entries/physicalism/. Accessed 30 October 2020.
Tuomela, R. (2013). *Social Ontology. Collective Intentionality and Group Agents.* Oxford: Oxford University Press.
Weintraub, E. R. (1979). *Microfoundations.* Cambridge: Cambridge University Press.
Ylikoski, P. (2014). "Rethinking Micro-Macro Relations". In: J. Zahle and F. Collin (eds.) *Rethinking the Individualism-Holism Debate. Essays in the Philosophy of Social Science.* Dordrecht: Springer, pp. 117–138.
Zahle, J. (2021). "Limits to Levels in The Methodological Individualism–Holism Debate", *Synthese* 198: 6435–6454.

CHAPTER 9

On the Nature of Money

> Sleepless, possessed, almost happy, I reflected that there is nothing less material than money, since any coin (a twenty-centavo piece, for instance) is, in truth, a panoply of all possible futures. Money is abstract, I said over and over, money is future time. It can be an evening just outside the city, or a Brahms melody, or maps, or chess, or coffee, or the words of Epictetus, which teach the contempt of gold; it is a Proteus more changeable than the Proteus of the Isle of Pharos. It is unforeseeable time, Bergsonian time, not the hard, solid time of Islam or the Portico. Adherents of determinism deny that in the world there is only one possible event, ed ist an event which could have happened; a coin symbolizes our free will.
>
> —Jorge Luis Borges, *The Zahir.*

The topic of the nature or essence of money has produced an abundant literature that has failed to come to a uniform conclusion. It has drawn the attention of famous philosophers, sociologists, and economists like Aristotle, Nicholas Oresme, David Hume, Adam Smith, Ferdinando Galiani, Georg Simmel, Max Weber, Carl Menger, Georg F. Knapp, John Maynard Keynes, Karl Polanyi, Ludwig von Mises, Friedrich von Hayek, and Joseph Schumpeter, among others. In the passage quoted above, Borges implicitly refers to the traditionally stated functions of

This is a modified version of my article 2021.

R. F. Crespo, *The Nature of the Economy*,
https://doi.org/10.1007/978-3-031-02453-5_9

money: unit of account, medium of exchange, and store of value. Furthermore, he stresses its abstract nature and saleability, linking it to a notion of time and even considering it a symbol of free will![1] The nature of money deserves to be thoroughly investigated: this chapter aims to do just that—as ambitious as it may seem…

The question about the nature of money has been associated with theories on its origin, about which we can only postulate hypotheses (Lawson 2016: 961). Although there are various theories on the origin of money, they can be grouped under two main positions: one positing a spontaneous evolution of money and the other one claiming that money is a product of intentional design. When considering money as a social institution, these two approaches relate to different corresponding conceptions of social institutions.

I believe that ascertaining the essence of money does not necessarily imply carrying out historical or archaeological studies to determine whether one thesis about the origin of money prevails over the other. As Simmel asserts at the beginning of his *Philosophy of Money*, the question about the essence of money is not historical but philosophical, while the question about the origin of money is not philosophical but historical (cf. 2004: 54). Here, I will not deal with the latter.

Referring to Aristotle's theory on money, Schumpeter states that, though presented in a "genetic form," it is "a theory in the ordinary sense of this term, that is to say, an attempt to explain what money is and what money does" (1954: 60). He also adds the following question and its answer, "Is it valid procedure to trace as far back as we can the history of an institution in order to discover its essential or its simplest meanings? Clearly not" (1954: 61).

In fact, natural experiments have demonstrated that both theories on the origin of money originated can coexist side by side. My country, Argentina, has an unfortunate history of monetary problems. On the upside, the country has become a real laboratory for economic practices. I have witnessed, as Georgina Gómez (2018, 2019) carefully describes, how different spontaneous, evolutionary, or internally regulated monetary systems and systems borrowed from other countries can actually work

[1] The American Institutionalist economist Wesley Mitchell explains how the introduction of money increases people's freedom (1944).

together at the same time.[2] This is not an abnormal phenomenon. Viviana Zelizer adopts a similar concept of multiple kinds of money based on the history of money in the United States between the 1870s and the 1930s. She provides several examples of monies and states (1994: 20–21):

> These objects have no common physical characteristic; they qualify as distinct monies because of the uses and meanings people assign to them, because of the distinctions they represent in everyday social life. Social monies certainly include officially issued coins and bills, but they also include all objects that have recognized, regularized exchange value in one social setting or another.

It is also worth noting that money has evolved over the years from its earliest forms—commodity monies with strong physical support—to the current fiat money and even more abstract forms of it. In short, we must look elsewhere to find the key to the nature of money. As Edith Stein has noted, "the essence of a reality isn't to be inferred from its development" ([1922] 2000: 238).[3]

Nonetheless, different positions on the origin of money grant different temporal or ontological priorities to its functions, thus bringing some of its essential features to light.[4] They either prioritize money as a medium of exchange or money as a unit of measure (fixed by an authority), roughly corresponding to their respective theories—spontaneous or designed

[2] See also Catan (2002). There is also the case of the use of the old Italian currency, the "lira" (replaced by the Euro) by the mafia for illicit transactions (https://www.bloomberg.com/news/articles/2019-06-15/italy-s-mafia-uses-the-old-lira-as-its-own-parallel-currency).

[3] See also Uskali Mäki (2009: 3, nt. 3). I have benefited greatly from reading this piece on the ontology of money, especially because it draws a clear distinction between money as universal and the specific instantiations of it or "currencies".

[4] Mäki (2009: 3) considers these functions as the essence of money. I prefer to say that they are essential characteristics of money, characterized by Aristotle as "eternal accidents" (*Metaphysics* V, 30 1025a 30–33) and called *proprios* by the scholastics.

origin.[5] The second position is also connected to the notion of money as a credit or a promise to pay (e.g., Ingham 2004; Fitzpatrick 2014).

In fact, money as an exchange instrument requires a measuring device. As Leonidas Zelmanovitz puts it, "money for us is a 'unit of account,' a 'standard of value,' *because* it is the preferred medium of exchange" (2016: 24). According to Menger (1892), this is so because money is the more saleable commodity, but the fact that money is a unit of account and a standard of value facilitates exchanges. Tony Lawson (2016: 967ff.) underscores money as a form of value, while, for J. M. Keynes (and Ingham 2018), "money of account, namely that in which debts and prices and general purchasing power are *expressed*, is the *primary* concept of a theory of money" (2013: 3, cursives in the original)—a view opposing Zelmanovitz'. It seems that these functions of money are correlative. Which comes first, the egg or the chicken? I will come back to this point.

Regardless of the evolution of money, some passages by Aristotle shed light on the meaning and nature of it, and I will look into them in this chapter. The first section will present Aristotle's thoughts on money, while the second section will analyze money as a unit of measurement, and the third section will deal with its conventional nature. In the fourth section, I will draw on two theories of signs to explain the nature of money as a sign, before offering a conclusion in the end.

Aristotle on Money

The first short passage of Aristotle's works that I will quote refers both to the nature of money and to its origin. In his *Nicomachean Ethics* (*NE*), he affirms:

> *Oion d'hypallagma tes chreias to nomisma gegone kata syntheken*. (*NE* V, 5 1133a 29–30)[6]

[5] Aristotle claims that both functions are interrelated. Oresme (1956: 4–5) asserts, "money does not directly relieve the necessities of life, but is an instrument artificially invented for the easier exchange of natural riches." David Hume ([1752] 1970: 32) posits, "Money is not, properly speaking, one of the subjects of commerce; but only the instrument which men have agreed upon to facilitate the exchange of one commodity for another."

[6] *Hypallage* means an exchange and *hypallasso* is the verb "to exchange." *Hypallagma* is "that which is exchanged" (Liddell & Scott 1900: 830). *Chreia* is a request of necessity,

The passage has been translated in several ways: "But money has become by convention a sort of representative of demand" (Ross); "money is the exchangeable representative of demand" (Liddell & Scott 1900: 830); "But demand has come to be conventionally represented by money" (Rackham); "And currency has become a sort of pledge of need, by convention" (Irwin). A French translation reads, "La monnaie est devenue, en vertu d'une convention, pour ainsi dire, un moyen d'échange pour ce qui nous fait défaut" (Jean Voilquin, Garnier, p. 219).[7] Schumpeter translates it as "money is a means of exchange [used] according to convention" (1954: 60). My translation is, "Money (*to nomisma*) has become (*gegone*) by convention (*kata syntheken*) like that which is exchanged (*Oion d'hypallagma*) for needs [or wants][8] (*tes chreias*)."

Aristotle states that exchange depends on equality. He offers the following example:

> Let A be a house, B ten minae, C a bed. A is half of B, is the house is worth five minae or equal to them; the bed, C, is a tenth of B; it is plain, then, how many beds are equal to a house, viz. five. That exchange took place thus before there was money is plain; for it makes no difference whether it is five beds that exchange for a house, or the money value of five beds. (1133b 24–29)

Consequently, for money to allow an exchange, it needs to have the value of the thing exchanged. This value has been assigned by convention. The following passage in Aristotle's *Nicomachean Ethics* explains that money can be this medium of exchange and that it additionally measures all things:

a need (Liddell & Scott 1900: 893). *Nomisma* means "money." *Kata syntheken* means "by convention.".

[7] Aristotelian scholar Carlo Natali wrote to me (April 21, 2020): "You quote many translations of Aristotle's *Nicomachean Ethics*; among the quotes I think Irwin's is the best, and Voilquin's is the least reliable. There is a problem about the translation of *chreia*, a Greek term that corresponds both to 'need' and to 'use'. Ross' translation as 'demand' is too modern. I think 'need' is a good translation in 1133a29-30." Scott Meikle (1995: 18) also thinks that it is more appropriate to translate *chreia* as need than as demand. Demand is a noun imbued of anachronistic economic connotations.

[8] A reviewer has suggested that the term "want" may be more adequate than "need," and I agree.

> This is why all things that are exchanged must be somehow comparable. It is for this end that money has been introduced (*eleluthen*), and it becomes in a sense an intermediate (*meson*); for it measures all things. (*NE* 1133a 17–21)[9]

Yet, he delves deeper:

> All goods must therefore be measured by some one thing, as we said before. Now this unit is in truth demand (*he chreia*), which holds all things together. (*NE* 1133a 25–27)[10]

However, needs or wants are qualitative properties that require a quantitative measure to allow exchange. Hence, Aristotle makes the previously quoted statement: "Money has become by convention like that which is exchanged for needs [or wants]." We conventionally assign a quantitative value to an instrument (money) and use an amount of it that is equal to the qualitative characteristic of the thing—its necessity—to acquire it ("the money value of five beds").

In Aristotle's time, money was commodity money. Referring to coins, he explains:

> At first their value was simply determined by their [metals] size and weight; but finally a stamp was imposed on the metal which, serving as a definite indication of the quantity, would save men the trouble of determining the value on each occasion. (*Politics* I, 9, 1257a 38–41)

Still, money value, Aristotle recognizes, is not intrinsic but conditional on its capacity to acquire the things needed: "for if those who use a currency give it up in favor of another, that currency is worthless, and useless for any of the necessary purposes of life" (*Politics* I, 9, 1257b 11). The value of money lies in its capacity to acquire what we need. Money has and symbolizes the value conferred by this capacity—its purchasing power—a value that equates the quality—need—and allows to fix a price in order to

[9] Ross' translation. Natali considers that *eleluthen* means "discovered" or "invented" rather than "introduced." This nuance is interesting because it suggests a convention about something that already existed.

[10] Ross' translation. Irwin translates *chreia* as need.

finally satisfy the need.[11] Uskali Mäki (2009: 4) speaks about the causal powers or dispositional properties of money, which are its functions.

As noted by Schumpeter (1954: 59), Aristotle remarkably enumerates the functions usually assigned to money.

1. It is a *unit of measurement*: "There must, then, be a unit, and that fixed by agreement [...] for it is this that makes all things commensurate, since all things are measured by money" (*NE* V, 5, 1133 b 19–23). In this sense, money is also a standard value.
2. It is a *means of exchange*: "all things that are exchanged must be somehow comparable. It is for this end that money has been introduced" (*NE* V, 5, 1133 a 19–21). "The various necessaries of life are not easily carried about, and hence men agreed to employ in their dealings with each other something" (*Politics* I, 9, 1257 a 33–35). "Money was intended to be used in exchange" (*Politics* I, 10, 1258 b 4).
3. It is a *value reserve*: "for the future exchange—that if we do not need a thing now we shall have it if ever we do need it—money is as it were our surety. Money is meant to be spent, but it can also be accumulated" (*NE* IV, 1, 1120 to 8–9). Thus, it becomes accumulated wealth, to the extent that it maintains its purchasing power. However, it should be underlined that this function of money—storage—only makes sense in terms of a "future exchange."
4. It has *purchasing power*: "for it must be possible for us to get what we want by bringing the money" (*NE* V, 5, 1133 b 11–13).
5. This feature also implies that money is a *means of payment*.[12]

These properties and capacities are correlative—indispensable for money to be money—and mutually supportive. The priority of a form of value or of a unit of measure lies at the core of a discussion between Lawson (2016, 2018) and Geoffrey Ingham (2018). According to Aristotle, they are simultaneous, mutually demanding essential characteristics of money.

[11] Ingham asserts that he understands money as "the power of abstract value that is *signified* by the *signifiers* that we call money" (2018: 847). Lawson views purchasing power as "money's power or potential to be used throughout a relevant economy to make purchases" (2018: 851).

[12] Ingham (2018: 841–842) distinguishes between medium of exchange and means of payment, as Aristotle does.

Ingham (2018: 845) speaks of simultaneity, and Lawson (2016: 970) also considers its possibility, but, if I understand their views correctly, these authors regard the ontological priority in different ways. In *Politics*, Aristotle states (I, 9, 1257a 35–37):

> The reason for (*pròs*) this institution of currency was that all the naturally necessary commodities were not easily portable; and men therefore agreed (*synéthento*), for the purposes (*pròs*) of their exchanges, to give and receive some commodity which itself belonged to the category of useful things and possessed the advantage of being easily handled for the purpose (*pròs*) of getting the necessities of life.

In *Politics* I, 9, Aristotle implicitly applies his doctrine of the four causes to particular monies[13]: there is a material cause (in his time, as mentioned, money was a commodity; today, this material cause can be digital), a formal cause (the value assigned to it, based on its purchasing power), an efficient cause (the people who agree to do that), and a final cause (as a medium of exchange, 1257b 1 and 1258b 4). These four causes are necessary for money to exist, but, for Aristotle the final cause has ontological priority. Indeed, it is safe to say that the medium of exchange function takes priority. This priority in not necessarily historical, but ontologically genetic. Yet, money cannot be a medium of exchange without owning or having been assigned a measure of value.

There are more famous passages about money in Aristotle's *Politics*. In *Politics* I, 9, 1257a 6–14, he introduces his explanation of the meaning of chrematistics by providing an example of the use of a shoe "for wearing or for exchange"; in the latter case, "in return for money or for food," thus "implicitly" considering a price or value of exchange for the shoe.[14]

Money differs from price: money measures a need through a price and an exchange value, enabling the exchange. It provides a measure of the

[13] Explained in Chapters 3 and 7. He distinguishes four real causes (efficient, formal, material, and final: *Metaphysics* I, 3–10; Physics II, 3) that provide different types of explanations—"a doctrine of four becauses" (Ackrill 1981: 36) that answers the following questions: who made this, why is this this thing and no other (the essence), out of what is this made, and for whose sake is this made?

[14] Exchange-value differs from price, because price is the expression of the former in monetary terms. Exchange value stems from the use value, given the availability of the object, and it is measured by the price determined through money. I say "implicitly" to pay attention to the risk of anachronism when using these notions.

need, thus establishing an exchange value expressed by the price. Money is not "operative"; without a price, it is "abstract": it is an open possibility or potentiality, "a panoply of all possible futures," as Borges nicely puts it. Aristotle asserts:

> [t]his is why all goods must have a price set on them (*tetimestai)*; for then there will always be exchange, and if so, association of man with man. Money, then, acting as a measure, makes goods commensurate and equates them; for neither would there have association if there were not exchange, nor exchange if there were not equality, nor equality if there were not commensurability. (*NE* V, 5 1133b 15–20)[15]

That is, on account of money, needs can be expressed in prices. In short, for Aristotle, money is a measurement instrument that makes it possible to set a price for an object. The price represents the exchange value of the object, which in turn expresses the need for the object sold or bought. In addition, the price remains "abstract"—though less than money—until the thing or service is effectively bought.

Also, money plays a role in social cohesion, as, by facilitating exchanges, it is an instrument to distribute *polis* members' needs: money facilitates exchanges in society and makes society possible—it is a condition for its existence. In this sense, it "accompanies" exchanges that emerge as a natural process. Aristotle holds that an exchange *arxamen kata physin* "arises by nature" (*Politics* I, 9, 1257a 15, my translation): "and is due to some men having more, and other less, than enough for their needs" (1257a 15–16). That is, money must be understood in the context of the whole Aristotelian social system. This is the reason why Aristotle deals with the topic of money in the first book of his *Politics*.

This Aristotelian causal analysis refers to different monies or currencies. However, Aristotle does not only refer specifically to money but, primarily, to its very concept, which is instantiated in specific monies or currencies. For Aristotle, money possesses the value of the commodity used as money, but it is conditioned to its acceptance; ultimately, this value is conferred by the mentioned powers or functions that finally act through specific instantiations. I think that Mäki's formulation expresses Aristotle's thought very accurately:

[15] Ross's translation. For Natali, at 1133b15, "have a price set on them" is too modern, *tetimestai* means in general "to be appreciated, estimated".

> Using the language of universals and particulars, we may characterize their relationships by saying that the money universal—moneyhood or moneyness— manifests itself in particular items of money and in generic currencies. We may also say that these relationships are characterized by multiple realizability: money has multiple realizations in specific currencies and particular items of money. The money universal is one, its realizations are many. (2009: 8)

He adds, "In the case of the money universal, existence and exercise are linked in that existence requires exercise: money's powers need to be exercised in order for them to exist [...] moneyhood only exists as instantiated in particular currencies and money items rather than independently of them" (2009: 11).

At this point, it may prove interesting to recall the classical doctrine of universals. Universals exist in singular objects as their essence (*universal in essendo*), in the concepts expressing this essence (*universal in repraesentando*), and in the words meaning the plurality of things that share the same "universal" characteristic (*universal in significando*). However, the most "real" universal is embodied in a specific thing: this animal, this house. The universal "law" or "rule" is instantiated in specific laws and rules. The universal "temperance" is instantiated in this temperate person or action. The same applies to money: its real condition is a specific currency—more so, a specific amount of money: a price paid, a payment, or an investment, etc. As Ann Davis notes, "money is only 'real' when traded for other commodities, including material objects as well as financial assets" (2017: 169).

The universal money—money as moneyness—is typically instantiated in currencies, which have powers that are exercised in specific, real dealings. That is, we can consider money in three ways: universal money, specific currencies, and specific monetary dealings that actualize their potentiality. The Aristotelian categories of potentiality (*dynamis*) and act (*energeia*) can also be applied here: universal money and particular currencies have a purchasing power or potentiality (capacity) that is actualized in specific transactions. In the case of exchanges, the exercise implies an exchange value and a price, which represents the value of a need (or want).

In the next sections, I will deal with money as unit of account, its conventional character, and money as a sign.

Money as a Unit of Measurement

For several authors, money is an abstract reality (see, e.g., Aglietta 2001; Fitzpatrick 2014: 209ff.). Indeed, "money," as noted earlier and concurring with Mäki, is a universal abstract term. Instead, currencies and prices are specific.

Once again, I will offer three quotes from Aristotle (already mentioned):

> This is why all things that are exchanged must be somehow comparable. It is for this end that money has been introduced, and it becomes in a sense an intermediate (*meson*); for it measures all things. (*NE* V, 5, 1133a 17–21)

> Money, then, acting as a measure, makes goods commensurate and equates them. (*NE* V, 5, 1133b 16–17)

> There must, then, be a unit, and that fixed by agreement [...] for it is this that makes all things commensurate, since all things are measured by money. (*NE* V, 5, 1133 b 19–23)

In other words, for Aristotle, money as a specific currency is a conventional unit of measurement. For Charles S. Peirce, it is also a means to measure value (cf. 1931–1958: 1.122 and 2.780). John Searle notes that "all you need to have money is a system of recorded numerical values whereby each person (or corporation, organization, etc.) has assigned to him or her a numerical figure which tells at any given point the amount of money they have, and they can then use this money to buy things" (2006: 64). This definition applies to instantiated conventional money.

We use figures to measure things. Aristotle formulated a related theory of mathematics and numbers. This theory has been recently readapted by the School of Sidney and other mathematicians who think that mathematical objects are "fictional entities grounded on physical objects" (Mendel 2004: 2)—universals *in significando* and *repraesentando*, grounded on universals *in essendo* or instantiated in specific objects. James Franklin, in his chapter on "Aristotelian Realism" in the *Philosophy of Mathematics* volume of the *Handbook of the Philosophy of Science* (Elsevier), maintains that mathematics studies the quantitative and structural aspects of things, like ratios, patterns, or relations. He views mathematicians as "scientists who study patterns or forms that arise in nature" (2008: 103). The idea of mathematics as a language can be included in this conception within a theory of signs. In this theory, mathematical symbols are "instrumental" and "conventional" signs of mathematical concepts, which are "formal" and "natural" signs of real structures or properties.[16]

Franklin (2008: 101–102) asserts,

> Aristotelian realism unifies mathematics and the other natural sciences. It explains in a straightforward way how babies come to mathematical knowledge through perceiving regularities, how mathematical universals like ratios, symmetries and continuities can be real and perceivable properties of physical and other objects, how new applied mathematical sciences like operations research and chaos theory have expanded the range of what mathematics studies, and how experimental evidence in mathematics leads to new knowledge.

For Aristotle, Jonathan Lear explains, mathematics is true, "because it accurately describes the structural properties and relations which actual physical objects do have" (1982: 191). This position parallels a realist view of measurement, which considers that the latter is based on quantitative attributes of reality (see, e.g., Joel Michel 2005). A number—for instance, 1, I, one, un, uno/a...—is a conventional sign for a notion—the concept of one per se—that is a formal and natural sign of a quantitative aspect of reality, the unity. One meter is a conventional measure of a specific length that is a quantitative aspect of a material element. To say that a table is one meter and 50 centimeters long is to say that the number 1.5 is a relation between the length of the table and the standard meter. This measurement can be almost exact.

Let us analyze the case of money. "One dollar," the nominal "face value" of an instantiated money or currency, is a conventional sign that measures its potential valuableness or purchasing power—a power that is exercised in specific economic dealings. In the case of exchanges, this is expressed by price: a quantity that represents the economic (exchange) value of a needed item. Price is the measure of the exchange value of objects (cf. Adam Smith [1776] 1981: 46).[17] Price as determined

[16] I will explore the meaning of instrumental, conventional, formal, and natural signs in the fourth section.

[17] Though, as already mentioned, money can be used as store of value or for other goals, the "phylogenetic" aim of it is ultimately to be used in exchanges: money is

by money is a conventional and instrumental quantitative sign of the economic value of the thing bought or sold, which is a natural and formal sign of the need for it.

A Conventional Sign

Aristotle upholds that money (as specific currency) is a representation *kata syntheken*—that is, "by convention." Aristotle uses the same expression to refer to languages: "a name is a spoken sound significant by convention—*kata syntheken*" (*On Interpretation* 16a 30–31; see also *Poetics* 1457a 10). Language is not by nature (*On Interpretation* 16a31). *Syntheke* (Liddell & Scott 1900: 766) means "a conventional agreement" and *kata syntheken*, "conventionally." However, this convention emerges naturally; dictionaries and language academies follow later. Dictionaries are continuously modified, and new words are constantly introduced when evidence shows their frequent use. Additionally, language is a means of expressing ideas, of communication; languages are specific instantiations of language, and they are conventional. Yet, language only exists in languages, and, therefore, convention is necessary for language to actually exist.

Likewise, Aristotle adds about money: "it exists not by nature" (*NE*, 1133 a 31). The existence of money depends on its instantiation, which is conventional. Money needs to be instantiated and exercised in order to become effective, just like the decimal system or the vocabulary of a language—all arbitrary signs that human beings can create using their rationality to express a reality: in the case of money, a numerical sign that represents an asset, something that the owner possesses. Karl Polanyi notes the similarity between money and language. He affirms, "Anthropologically, money should be defined as a semantic system, broadly similar to language, writing, or weights and measures." He describes the functions of money as "uses," and he states that "'money uses' represent the 'purpose' of money as a semantic system" (1977: 97). Thus, as asserted, money as a sign supervenes on these uses and capacities.

According to this meaning of instantiated money, it could thus follow that money is a social construct, whose nature is independent of its origin. It is probably essential that fiat money be created by the state, but this is not essential for money itself. In conclusion, an essential characteristic

essentially aimed at facilitating exchange; as Aristotle posits, this, as argued, is the final cause of money according to Aristotle.

of money is to be a conventional device, regardless of its spontaneous or designed origin. As Ann Davis holds, the functions of money "are assigned by convention and often [though not necessarily, I add] enforced by government" (2017: 23).[18] David Colander clearly sentences, "the state is not necessarily involved with the essence of money [...] Money is a creature of society, not of the state" (2019: 64). There is abundant literature arguing that money is a social institution. I agree, as far as a generally shared convention—regardless of its origin and "legal" status—is considered a social institution.[19] The specific instantiated money or currency is a social institution in this broad sense.

Money is important for its measuring ability—much like a ruler or a meter stick—serving as a conventional measure to enable comparisons. Its stability is important—Aristotle points out that, even though money can change in value, it is quite stable (*NE*, 1133 b, 14–15)—because stability allows for relative measurements, setting prices, and making exchanges. With money, exchange values can be expressed in monetary terms. The conventional character of money makes trust—in addition to an adequate monetary policy—a fundamental factor for its stability. As Simmel asserts, "money transactions would collapse without trust" (2004: 179).

What do money and prices represent? They stand for the value of needs—*chreia*—which bring about exchanges. The fact that something is more necessary or scarcer than something else is reflected in the fact that more money is required to obtain it. That is, as already noted, money conventionally expresses a quality quantitatively through prices.

[18] As Smit et al. (2011: 16) note, commodity and fiat money are backed by the same thing: human preferences.

[19] I believe it is not necessary to either adopt or reject any of the different views on which social institutions are conceived or the way in which money is considered. John R. Searle has repeatedly argued that money is a social institution (1995, 1998: 126–128, 2005). However, my definition of institution differs from Searle's. For him, we need a constitutive rule to have an institution. See also Ann E. Davis (2017). For discussions on Searle's position, among others, see Barry Smith (Smith and Searle 2003), Mäki (2009), J.P Smit, Filip Buekens and Stan du Plessis (2011), Georgios Papadopoulos (2015), and Francesco Guala (2016a, b). About the different notions of social institutions, see Seumas Miller (2019). Viviana A. Zelizer (1994) considers money as embedded in social systems. I think that Geoffrey Hodgson's concept of institution is wider than Searle's and compatible with the notion of money supported here. He states, "systems of language, money, law, weights and measures, traffic conventions, table manners, and all organizations are institutions" (2015: 501; see also 2006).

Simmel stresses this fact affirming, "one of the major tendencies of life—the reduction of quality to quantity—achieves its highest and uniquely perfect representation in money" (2004: 280).

Money as a Sign

Revisiting language, it is instantiated in a specific system of signs: a conventionally organized set of *significant sounds* (cf. Aristotle, *On Interpretation* 16a 30–31), as considered by contemporary linguistics and semiotics. In this section, I will argue that these theories apply to money as a sign, offering an updated theoretical framework to Aristotle's views on it.

However, to start this analysis, I will use a conception of sign as presented by an old scholar, John Poinsot, in his theory of signs—a notion that I have often found useful. Poinsot (1589–1644) was a Portuguese thinker who wrote *Tractatus de signis* (*Treatise on Signs,* [1631–35] 1985) and stood as a foremost representative of the so-called *Conimbricenses* (from Coimbra, Portugal), a school of seventeenth-century logicians that came up with a highly elaborated theory of signs. It should be noted that a very rich semiotic theory was developed in medieval times (see S. Meier-Oeser 2003). For Poinsot, a sign is "that which represents something other than itself to a cognitive power" ([1631–35] 1985, p. 25), and he classifies signs accordingly ([1631–35] 1985, p. 27):

> [I]nsofar as signs are ordered to a [knowing] power, they are divided into formal and instrumental signs; but insofar as signs are ordered to something signified, they are divided according to the cause of that ordering into natural and stipulative [or conventional] and customary. A formal sign is the formal awareness which represents of itself, not by means of another. An instrumental sign is one that represents something other than itself from a pre-existing cognition of itself as an object, as the footprint of an ox represents an ox. And this definition is usually given for signs generally. A natural sign is one that represents from the nature of a thing, independently of any stipulation and custom whatever, and so it represents the same for all, as smoke signifies a fire burning. A stipulated sign is one that represents something owing to an imposition by the will of a community, like the linguistic expression "man." A customary sign is one that represents from use alone without any public imposition, as napkins upon the table signify a meal.

Table 9.1 John Poinsot's classification of signs

In respect to the thing signified	In respect to the knowing power
Natural: smoke, concept of fire, value	Formal: the concept, value
Conventional: "fire," nominal face value, Price	Instrumental: nominal face value, price

Table 9.1 clarifies this classification.

Based on Poinsot's conception, words (like "fire") are considered instrumental and customary or conventional signs, while concepts (like the notion of fire) are formal and natural signs of the apprehended reality. Poinsot's formal sign directly points to a real reference. Aristotle mentions both kinds of signs—natural and conventional—in *On Interpretation* (16a 4–7). According to this typology, price is a conventional and instrumental sign of the exchange value of the thing needed determined by money. The exchange value is a natural and formal sign of the need of it when the transaction is effectively done. Currencies' nominal face values are instrumental and conventional signs of their values on account of their purchasing power, and realized by prices that are conventional and instrumental signs of the exchange value, a natural sign of the need.

I will now turn to a more contemporary author viewed as the founder of modern semiotics, Charles S. Peirce (1839–1914), to complement Poinsot's theory. In addition to his interest in economic affairs and economic science (see James R. Wible 2008; Richard Swedberg 2011), Peirce formulated a theory of signs that is also applicable to money. He states:

> A sign, or *representamen*, is something which stands to somebody for something in some respect or capacity. It addresses somebody, that is, creates in the mind of that person an equivalent sign, or perhaps a more developed sign. That sign which it creates I call the *interpretant* of the first sign. The sign stands for something, its object. It stands for that *object*, not in all respects, but in reference to a sort of idea, which I have sometimes called the ground of the *representamen*. (CP 2.228, my italics)

Namely, there are three elements: the *representamen* (the sign, as the word, or an amount of money, or a price), the *interpretant* (the sign in the interpreter's or cognoscente's mind: my concept or idea of the amount of money or price), and the real *object* (the thing signified by the word and the concept; the need for the thing). That is, the sign or *representamen* is a means between the object and the *interpretant*. It also provides a plus, something else, in relation to the object. Peirce affirms:

> To stand for, that is, to be in such a relation to another that for certain purposes it is treated by some mind as if it were that other. Thus a spokesman, deputy, attorney, agent, vicar, diagram, symptom, counter, description, concept, premise, testimony, all represent *something else*, in their several ways, to minds who consider them in that way. (CP 2.273, my italics)

There is "something else," an interpretation of the object, resulting from the sign. Returning to Poinsot's classification, the relation between the sign (*representamen*) and the *object* is conventional and instrumental; the relation between the *representant* and the *object* is natural and formal; and the relation between the sign and the *representant* is partly natural and partly conventional.[20]

Additionally, Peirce distinguishes three types of signs:

> One very important triad is this: it has been found that there are three kinds of signs which are all indispensable in all reasoning; the first is the diagrammatic sign or icon, which exhibits a similarity or analogy to the subject of discourse; the second is the index, which like a pronoun demonstrative or relative, forces the attention to the particular object intended without describing it; the third [or symbol] is the general name or description which signifies its object by means of an association of ideas or habitual connection between the name and the character signified. (CP 1.369)

Language and money belong to the third type of signs: there is no similarity nor physical connection—just a convention. He explains: "The objects of the understanding, considered as representations, are symbols, that is, signs which are at least potentially general" (CP: 1.559).

Concerning the sign interpretation (the *interpretant*), third-type signs are influenced by personal and social elements. Although words have a socially recognized meaning, they can also resonate differently with different people in different situations or contexts. Similarly, a price

[20] On the relation between Peirce and Poinsot, see M. Beuchot and J. Deeley (1995).

has different interpretations according to an individual's specific situation (*interpretant*). As Marshall holds, referring to prices, "the same price measures different satisfactions even to persons of equal incomes" ([1890–1920] 1962: 15). Also, he affirms, "the significance of a given price is greater for the poor than the rich" ([1890–1920] 1962: 16). Likewise, though basing his reflections on a different tradition (Paul Ricoeur), Richard Ebeling (1990: 187) refers to prices as follows:

> By themselves, prices are nothing more than quantitative ratios—magnitudes of things. They become more than this when viewed from within the setting in which they arise, and they arise from the actions of men. The intentionalities of men as reflected in prices are what require interpretation, and not the prices themselves.

This assumes the existence of a shared world that enables individuals to understand themselves. The role of entrepreneurial alertness in entrepreneurial discovery, leading to a tendency toward equilibrium in the market process, has been highlighted by Israel Kirzner (e.g., 2000). Individual interpretations in specific situations must also be borne in mind. Dave Elder-Vass notes:

> The value of a thing is not an objective quality, but rather the price that it ought to exchange at, and this is a subjective quantity in the sense that different individuals may take different views on it, but also a socially constructed quantity in the sense that in doing so they draw on socially or intersubjectively shared lay theories of value. [...] Many factors contribute to the determination of prices, but no price is viable unless it can be justified by the social actors involved. There is, in other words, no price without value. (2019: 1497)

Dan Fitzpatrick similarly affirms, "economic value is a subjective matter which then becomes an intersubjective matter through the process of exchange" (2014: 12 and see 49). In short, we have a *representamen*, money and prices—money measuring exchange value—and a *representant*—our idea of price—that refers to the object valued because it is needed. In addition, adopting Peirce's theory of signs allows for the notion that the symbolic character of money does not automatically imply its neutrality, as Ingham (2004: 60) complains.

Summing up, these theories of signs can be applied to money. There are natural and instrumental sign relations: the exchange value is a natural sign of the need, while the price is an instrumental sign of the exchange value. Nominal face values are instrumental signs of their values on account of their purchasing power. Peirce's *representamen* can be considered as the nominal face value, and the price, the *interpretant* as my concept or idea of the amount of money or the price, and the real *object*, the exchange value, and the need for the thing.

Conclusion

Aristotle's passages on money shed light on its nature. Building on his ideas, I have distinguished three levels of understanding money: (1) universal money or *moneyness*; (2) money as instantiated in particular currencies with nominal face values (money of account) that measure this potential purchasing value; and (3) currency in exercise that actualizes money/currency potentialities through specific prices, signs of exchange values, and needs/wants. The fact that the nominal face value of money means *potential* purchasing value shows that this value can actually change when exercised.[21] Currencies are performative signs, in the sense that they have a potentiality or purchasing power.

In the introduction of this chapter, referring to the priority of money of account or of money as a form of value (and a medium of exchange), I asked, "which comes first, the egg or the chicken?" Then, in the first section, I quoted this passage by Aristotle: "Money has become by convention *like that* which is exchanged for needs [or wants]" (italics added). From this and subsequent quotations from Aristotle recognizing all the correlative functions of money, I concluded that money as currency is a sign of a potential purchasing value that is actualized in specific transactions. In other words, as noted by Keynes, the money of account takes priority. However, money as a sign is present at the second level, while its purchasing power—only potential at the previous levels—actually materializes at the third level. The question is, what comes first: the sign (specific instantiations of universal money in currencies) or the signified (the exercise of the purchasing power)? It is tempting to go with the signified, but money is a performative sign with a real potentiality. For Aristotle, it seems that the priority depends on the level of consideration. Yet, an

[21] Keynes states, "money of account is the *description* or *title* and the money is the *thing* which answers to the description [...] if the thing can change, whilst the description remains the same, then the distinction can be highly significant" (2013: 3).

ontological question remains, what comes first, potentiality or action? In addition, universal money is instantiated in currencies for the purpose of exchange, a function that as final cause is ontologically prior to the other functions of currencies. Ultimately, the definition and ontological priority of money and its functions depends on which level of money we are considering. Still, it is clear that the three levels are accidental. The substantial character of coins and bills is irrelevant for the essence of money.

A final remark. Aristotle stresses the relevance of money's value stability because "money is as it were our surety," but money—like other goods—does not always have the same value (*NE* V, 5, 1133b 14–15). Aristotle considers this possibility even when his notion hinged on commodity money. Today that money is fiduciary money: its value depends on people's trust in it. The sign of the loss of trust in money is its increased circulation velocity. This is a highly subjective and unstable factor that may have a great impact.

Argentina proves a good case study because it has suffered periods of high inflation, including two hyperinflations, and periods of low inflation rates. For example, from February 2018 to February 2019, Argentina's inflation rate stood at 51%. It would have been 20% if the speed of circulation had remained stable (see https://ucema.edu.ar/cea/postal-velocidad). Emiliano Basco, Laura D'Amato, and Lorena Garegnani (2006) studied the relation between inflation, monetary expansion, and speed of money circulation in Argentina from 1975 to 2005. The correlation between inflation and velocity is clear (see Fig. 4.2, p. 15). The reasons are clear: people feel more uncertain amidst political instability or flawed policies and look for certainty, fleeing from the domestic currency to trustable assets: people try to avoid the inflationary tax by getting rid of the money as soon as possible. This behavior produces a vicious cycle: escaping from domestic money leads to increased velocity of circulation, which in turn causes higher inflation. Individual and social psychology bears a strong impact on the value of money. These issues cannot be managed with economic means; the return of trust depends on political factors. Regarding the topic of this book, the conclusion here is that even monetary policies—at least partially—depend on non-economic factors and are contingent.

References

Ackrill, J. L. (1981). *Aristotle the Philosopher*. Oxford: Clarendon Press.

Aglietta, M. (2001). "Whence and Whither Money?" In: *The Future of Money*. Paris: OECD, pp. 31–72.

Basco, E., L. D'Amato and L. Garegnani (2006). "Understanding the Money—Prices Relationship Under Low and High Inflation Regimes: Argentina 1970–2005", Working Paper 2006/13, Banco Central de la República Argentina. http://www.bcra.gob.ar/pdfs/investigaciones/wp%202006%2013_i.pdf.

Beuchot, M. and J. Deely (1995). "Common Sources for the Semiotic of Charles Peirce and John Poinsot", *The Review of Metaphysics* 48/3: 539–566.

Catan, T. (2002). "Argentines Snowed Under by paper IOU's. Pesos or Pacificos? A Dizzying Array of Currencies Now Fill Up the Tills", *Financial Times* 11 April 2002: 4.

Colander, D. (2019). "Are Modern Monetary Theory's Lies 'Plausible Lies'?", *Real-world Economics Review* 89: 62–71.

Crespo, R. F. (2021). "On Money as a Conventional Sign: Revisiting Aristotle's Conception of Money", *Journal of Institutional Economics* 17/3: 459–471.

Davis, A. E. (2017). *Money as a Social Institution. The Institutional Development of Capitalism*. Routledge: Abingdon.

Ebeling, R. M. (1990). "What Is a Price? Explanation and Understanding". In: D. Lavoie (ed.) *Economics and Hermeneutics*. London and New York: Routledge, pp. 177–194.

Elder-Vass, D. (2019). "No Price Without Value: Towards a Theory of Value and Price", *Cambridge Journal of Economics* 43: 1485–1498.

Fitzpatrick, D. (2014). *Economic Reality: The Ontology of Money and Other Economic Phenomena*, PhD Thesis, Department of Philosophy, Logic and Scientific Method. London School of Economics. UMI Dissertation Publishing. UMI Number U198904. http://etheses.lse.ac.uk/1806/1/U198904.pdf.

Franklin, J. (2008). "Aristotelian Realism". In: A. Irvine (ed.) *Philosophy of Mathematics*, D. Gabbay, P. Thagard and J. Woods (general eds.) *Handbook of the Philosophy of Science*. Elsevier, pp. 149–150. http://web.maths.unsw.edu.au/~jim/irv.pdf.

Gómez, G. M. (2018). "Why Do People Want Currency?", *Evolutionary and Institutional Economics Review* 15: 413–430.

Gómez, G. M. (2019). "Money as an Institution: Rule versus Evolved Practice? Analysis of Multiple Currencies in Argentina", *Journal of Risk and Financial Management* 12: 1–13.

Guala, F. (2016a). "Philosophy of the Social Sciences: Naturalism and Anti-naturalism". In: P. Humphreys (ed.) *The Oxford Handbook of Philosophy of Science*. Oxford: Oxford University Press, pp. 43–64.

Guala, F. (2016b). *Understanding Institutions: The Science and Philosophy of Living Together*. Princeton: Princeton University Press.
Hodgson, G. M. (2006). "What Are Institutions?", *Journal of Economic Issues* XL/1: 1–25.
Hodgson, G. M. (2015). "On Defining Institutions: Rules Versus Equilibria", *Journal of Institutional Economics* 11/3: 497–505.
Hume, D. ([1752] 1970). "Of Money". In *David Hume Writings on Economics*, edited with an Introduction by Eugene Rotwein. Madison: The University of Wisconsin Press, pp. 33–46.
Ingham, G. (2004). *The Nature of Money*. Cambridge: Polity Press.
Ingham, G. (2018). "A critique of Lawson's 'Social Positioning and the Nature of Money'", *Cambridge Journal of Economics* 42: 837–850.
Keynes, J. M. (2013). *A Treatise on Money*. Cambridge: Cambridge University Press.
Kirzner, I. M. (2000). *The Driving Force of the Market*. London: Routledge.
Lawson, T. (2016). "Social Positioning and the Nature of Money", *Cambridge Journal of Economics* 40: 961–996.
Lawson, T. (2018). "The Constitution and Nature of Money", *Cambridge Journal of Economics* 42: 851–873.
Lear, J. (1982). "Aristotle's Philosophy of Mathematics", *The Philosophical Review* 91: 161–192.
Liddell, H. G. and R. Scott (1900). *An Intermediate Greek-English Lexicon*, Founded upon the 7th ed. of Liddell and Scott's *Greek-English Lexicon*, 1889.
Mäki, U. (2009). 'Reflections on the Ontology of Money', paper presented at the Vancouver workshop on The History and Philosophy of Money, 13–14 November 2004, and at the ISNIE Conference in Barcelona, 22–24 September 2005.
Marshall, A. ([1890–1920] 1962). *Principles of Economics* (8th edition, 1920). London: MacMillan.
Meier-Oeser, S. (2003). "Medieval Semiotics". In: E. Zalta (ed.) *Stanford Enciclopedia of Philosophy*. http://plato.stanford.edu/entries/semiotics-medieval/. Accessed 22 September 2015.
Meikle, S. (1995). *Aristotle's Economic Thought*. Oxford: Clarendon Press.
Mendel, H. (2004). "Aristotle and Mathematics".In E. Zalta (ed.) *Stanford Encyclopedia of Philosophy*. http://plato.stanford.edu/entries/aristotle-mathematics/. Accessed 5 January 2020.
Menger, C. (1892). "On the Origin of Money", *The Economic Journal* 2: 239–255.
Michel, J. (2005). "The Logic of Measurment: A Realist Overview", *Measurement* 38: 285–294.

Miller, S. (2019). "Social Institutions". In E. Zalta (ed.) *Stanford Encyclopedia of Philosophy*, on line http://plato.stanford.edu/entries/social-institutions/. Accessed 20 January 2020.

Mitchell, W. C. (1944). "The Role of Money in Economic History", *Journal of Economic History*, 4/Supplement: 61–67.

Oresme, N. (*circa* 1355). *De Moneta* (*On Money*). The DE Moneta of Nicholas Oresme and English Mint Documents, translated from the Latin with Introduction and Notes by Charles Johnson. Thomas Nelson and Sons: Edinburgh, 1956. https://mises-media.s3.amazonaws.com/The%20De%20Moneta%20of%20Nicholas%20Oresme%20and%20English%20Mint%20Documents_2.pdf.

Papadopoulos, G. (2015). "Collective Intentionality and the State Theory of Money", *Erasmus Journal for Philosophy and Economics* 8: 1–20.

Parsons, T. (1951). *The Social System*. Glencoe: The Free Press.

Peirce, C. S. (1931–1958). *Collected Papers* Vols. 1–8, C. Hartshorne, P. Weiss, A. Burks (eds.). Cambridge, MA: The Belknap Press of Harvard University.

Poinsot, J. [1631–1635] 1985. *Tractatus de Signis. The Semiotic of John Poinsot*, edited by John N. Deely. London, Berkeley and Los Angeles, CA: University of California Press.

Polanyi, K. (1977). *The Livelihood of Man*, edited by Harry W. Pearson. New York: Academic Press.

Schumpeter, J. A. (1954). *History of Economic Analysis*. London: Allen & Unwin.

Searle, J. R. (1995). *The Construction of Social Reality*. New York: The Free Press.

Searle, J. R. (1998). *Mind, Language and Society*. New York: Basic Books.

Searle, J. R. (2005). "What Is an Institution?", *Journal of Institutional Economics* 1: 1–22.

Searle, J. R. (2006). "Social Ontology: Some Basic Principles", *Papers* 80: 51–71.

Simmel, G. ([1900] 2004). *The Philosophy on Money*. Third enlarged edition, edited by David Frisby. Translated by Tom Bottomore and David Frisby. London: Routledge.

Smit, J. P., F. Buekens and S. du Plessis (2011). "What Is Money? An Alternative to Searle's Institutional Facts", *Economics and Philosophy* 27: 1–22.

Smith, A. ([1776] 1981). *An Inquiry into the Nature and Causes of the Wealth of Nations*. Indianapolis: Liberty Fund.

Smith, B. and J. R. Searle (2003). "The Construction of Social Reality: An Exchange", *American Journal of Economics and Sociology* 62: 285–309.

Stein, E. ([1922] 2000). *Philosophy of Psychology and the Humanities*, Vol. 7 of the Collected Works of Edith Stein, edited by Marianne Sawicki. Translated by Mary Catharine Baseheart and Marianne Sawicki, Washington, DC: ICS Publications.

Swedberg, R. (2011). "The Role of Senses and Signs in the Economy: More on the Centrality of Materiality", *Journal of Cultural Economy* 4: 423–437.
Wible, J. R. (2008). "The Economic Mind of Charles Sanders Peirce", *Contemporary Pragmatism* 5: 39–67.
Zelizer, V. A. (1994). *The Social Meaning of Money.* New York: Basic Books.
Zelmanovitz, L. (2016). *The Ontology and Function of Money.* New York: Lexington Books.

CHAPTER 10

Economic Sciences

The promise made in the Introduction of this book was to analyze the nature and method of economic sciences based on the conclusions on the nature of the constituents of the economy.

Chapter 4 about the economy "in general" addressed economic actions. The conclusion was that the economic realm consists of accidental entities or properties that cannot be separated from underlying variable structures, capacities, and institutions, or from the agents that have diverse motivations influencing economic aspects. We face a contingent realm that calls for local analyses. The more local or particular the subject is, the truer the conclusions will be. However, there is a trade-off between this truth and the applicability of the results to other cases. The focus of the investigation must be defined according to its aims.

Chapter 5, focusing on the identity of the economic agent, pointed to the reductive character of this notion of standard economics: the economic agent is not only a "maximizer." We cannot isolate rational instrumental rationality from the other motivations of human social actions considered by Weber—value-rational, affective, and traditional reasons. This view of personal identity leaves ample room for contingency—every human person is different. With this view, one may wonder whether economic science is possible. Indeed, as mentioned in the previous paragraph, there is a trade-off between the realism of specific cases and the quest for generality in science. We need to strike a balance

R. F. Crespo, *The Nature of the Economy*,
https://doi.org/10.1007/978-3-031-02453-5_10

that depends on the subject and the aims of the investigation. There are grounds for generalizations thanks to the natural conditions and to people's social, psychological, and cultural common traits.

Exploring the economic agent as worker, Chapter 6 highlighted the relevance of considering the subjective dimension of labor, examining the philosophical grounds of this perspective. It also stated that the intersection of managerial and psychological studies shows the advantages of pursuing a calling for individuals, working groups, and entire corporations. Alas, concerned with work as a mere factor of production, current economics neglects its most valuable elements almost entirely. The conclusion is that economics should re-think its treatment of work. The identity of the economic agent—a human person with a plethora of motivations seeking personal fulfillment and a sense of purpose in life—has an impact on labor that economics cannot but seriously consider.

Chapter 8 stressed the metaphysical accidental character of macroeconomic entities. The macroeconomic stuff is a system that consists of interactions between individuals and their environment, featuring properties and conducts that cannot be reduced to an aggregation of individual conducts. These new elements have a strong relationship with cultural, social, and political factors and make the behavior of the macroeconomy highly context-dependent. Finally, Chapter 9 about fiduciary money also underscored the contingent nature of its value, greatly depending on people's trust in it.

All these characteristics of its subject matter link economics closely to the social, political, cultural, historical, and geographical dimensions where it—micro and macroeconomy—unfolds. This nature of economics and its method will surface more clearly in the rest of this chapter.

In my book published in 2013, I wrote a chapter (3) about economic science, where I proposed two central conceptions of economic science, corresponding to the division of the economy outlined in Chapter 4—the broad and the strict or focused notions of the economy. Then, in my book published in 2020, I also wrote a chapter on economic science, but I used the plural, "economic sciences," and I subsumed the two conceptions introduced in the previous book under one of the economic sciences, "economic theory." The other economic sciences considered in the 2020 book were history and statistics, normative economics, and "the art of economics." The reason for this distinction is that different levels in economic reality or in its analysis call for different scientific approaches. I found sources for these later distinctions in the views on economics crafted by John Stuart Mill, John Neville Keynes, and Carl Menger. In this chapter, I will summarize the notions of the previous books before introducing the ideas coined by Aristotle, Cartwright, Lawson, and Mäki on economic science. Finally, I will try to integrate all these views to show how they meet the requirements of the mentioned constituents of the subject matter of economics.

Previous Ideas on the Nature and Method of Economic Sciences

In his classical book *The Economic Point of View* 1976, Israel Kirzner introduced at least six different groups of definitions of Economics. Kirzner points out that "the depressing lack of unanimity among these formulations has led writers to doubt seriously whether they have any value at all" (1976: 5). In fact, most economists do not worry about the meaning of economics and are guided by pragmatic considerations, as Backhouse and Medema remark (2009: 231): they just do economics. These authors also interestingly note that one possible reason for this dispersion is the excessively broad character of the subject of economics (2009: 223).

Though recognizing that great plurality of views, Kirzner identifies two prevailing groups of definitions that were previously characterized by Lindley M. Fraser (1937):

> Fraser has classified definitions of economics into Type A definitions and Type B definitions. Type A definitions consider economics as investigating a particular department of affairs, while Type B definitions see it as concerned with a particular aspect of affairs in general. The specific department singled out by Type A definitions has usually been wealth or material welfare. The aspect referred to in Type B definitions is the constraint that social phenomena uniformly reveal in the necessity to reconcile numerous conflicting ends under the shadow of an inescapable scarcity of means. (1976: 17)

Several economists like Ronald Coase (1978: 206ff.), Lionel Robbins (calling them "material" and "scarcity" definitions, 1935: Chapter 1), Ernest H. Phelps Brown (calling them "field" and "discipline-determined," 1972: 7), Karl Polanyi ("substantive" and "formal" meaning, 1971: 139–140), and Dani Rodrik (2015: 7) share, with their nuances, this classification, dealing with the two concepts of economy formulated in Chapter 4. Kirzner states, "Our problem, then, relates not to the scope of the subject 'economics,' but to 'economic theory'" (1976: 4), implicitly recognizing that economics is broader than economic theory. This is what I think. In my 2020 book, drawing from Mill, Menger, and Neville Keynes, I broadened the scope of economic science, not only considering the previous versions of economics ("economic theory") but also, as noted, history and statistics, normative economics, and "the art of economics."

For Mill, political economy deals with human beings' behavior in society "solely as a being who desires to possess wealth, and who is capable of judging the comparative efficacy of means for obtaining that end" ([1844] 2006: 321). However, as already noted in Chapters 2 and 5, he views this definition as a simplifying abstraction. Consequently, he points to the need of developing an art of political economy formulated by "practical men." In addition, Mill considers two kinds of arts: the art of definition of ends (which concerns morality) and the art of performing the actions directed to these ends, with the assistance of science (*System of Logic* 1882: 653; VI, XXI, 2) and the art concerning with ends (1882: 657). That is, for Mill we have economics, its normative art, and the art of achieving the ends indicated by the former.

Menger, in his article "Toward a Systematic Classification of the Economic Sciences" (1960, English translation of "Grundzüge einer Klassifikation der Wirtschaftswissenschaften," 1889), uses the plural term "economic sciences," and he proposes the "historical sciences of economics" (1960: 14) as the first of them: economic statistics and economic history, which investigate specific economic phenomena offering valuable inputs to economic theory.

For Menger, economic theory has a demonstrating (*Darstellung*) and understanding (*Verständnis*) role (1889: 6; 1960: 7). *Verstehen* and *Verständnis*, "to understand" and "understanding," refer to a particular form of explanation of the human sciences that tries to capture the intention of human actions. Menger also considers *practical or applied economics* (1960: 16, 21–22), including economic policy and the science of finance. For him, understanding is also a function of applied science (1960: 20), and he uses the term "practical sciences" (*praktische Wissenschaften*) as equivalent to "applied science," referring to *anthropina philosophia* (an expression coined by Aristotle, "*all* the sciences of man,"

Menger explains) which *reasonably* (*verständig*) apply general principles to specific cases (1889: 18, footnote note 1, 1960: 35, endnote 14).

Neville Keynes thinks that the premises of political economy are hypothetical, because he notes that there is a plurality of motives for economic actions aside from the purely economic ones ([1890] 1955: 16). This plurality is included in the "widest sense" of "political economy" or "economic inquiry" ([1890] 1955: 36 and 61). It involves different "departments," "inquiries," or "subdivisions" ([1890] 1955: 30, 34, 35, 61) with dissimilar methods depending on the nature of their subjects: "*economic uniformities, economic ideals and economic precepts*" ([1890] 1955: 31, 35). Accordingly, he proposes this distinction and definition:

> a *positive science* may be defined as a body of systematized knowledge concerning what is; a *normative or regulative science* as a body of systematized knowledge relating to criteria of what ought to be, and concerned therefore with the ideal as distinguished from the actual; and *art* as a system of rules for the attainment of a given end. ([1890] 1955: 34–35, all Keynes' italics)

For Keynes, art is a science ([1890] 1955: 118). He bases this thesis on Adam Smith's use of the term science referring to a systematic body of knowledge of theoretical propositions as well as practical rules ([1890] 1955: 35, nt 2).

Merging all these classifications, the following branches of "political economy in its widest sense" (Neville Keynes) or "economic sciences in general" (Menger 1960: 3) can be distinguished:

1. The historical sciences of economy: economic history and statistics.
2. Economic theory dealing with the economy in its broad and narrow senses. Positive economics deals with the economy in its restricted sense. In addition, in my 2020 book I proposed to introduce a scientific treatment of the actual ends of economic actions into economic theory, thus incorporating the broad conception of the economy as part of its subject matter. This is the Mengerian *Verständnis* aspect of economic theory.
3. Mill's normative economics, that is, his Teleology or Doctrine of Ends concerning economic goals, or Neville Keynes's normative economics.

4. The practical economic sciences, including economic policy and the science of finance according to Menger, Mill's art of economic practice, or Neville Keynes's applied economics.

These four economic disciplines feature different aims and characteristics:

1. The aim of historical sciences is *to describe* economic facts and provide information to economic theory. History is more relevant to human and social sciences than to exact and natural sciences. We do not need to know Pythagoras's times and circumstances to apply the Pythagorean theorem, but it is important to know the history of economic facts and thought to understand its concepts and developments. As for data, it is an essential component given the context-dependent nature of economics. Marcel Boumans (2015: ix) notes the characteristics of measurement in "field science" (sciences that are based on data that do not come from "clean and controlled environments"): context-dependence, sensitivity to local conditions, inexactness (Boumans 2015: 174). Measurement requires a "considered judgment" (Elgin 1996), like clinical practice (Boumans 2015: Chapter 5).
2. The aim of economic theory is *to explain* and *understand* economic phenomena. As already mentioned, I have distinguished positive economics from economic theory dealing with ends. Positive economics is standard contemporary economics dealing with means to attain given ends. It is concerned with identifying "efficient" or "pushing" causes of economic events, thus explaining them. It deals with a contingent subject matter and uses inductive, abductive, and deductive methods of reasoning.[1] The first two methods are not infallible. Positive economics uses models that have to be adapted to meet the specific conditions cited in their *explanandas,* as J. Maynard Keynes explains:

> It seems to me that economics is a branch of logic; a way of thinking [...] one cannot get very far except by devising new and improved models. This requires, as you say, 'a vigilant observation of the actual working of our system'. Progress in economics consists almost

[1] On abduction and economics, see Tohmé and Crespo (2013).

> entirely in a progressive improvement in the choice of models [...] Economics is a science of thinking in terms of models joined to the art of choosing models which are relevant to the contemporary world. [...] The object of a model is to segregate the semi- permanent or relatively constant factors from those which are transitory or fluctuating so as to develop a logical way of thinking about the latter, and of understanding the time sequences to which they give rise in particular cases. (Keynes 1973: 296, letter to Harrod, 4 July 1938).

This idea has been shared by Dani Rodrik. He holds that economists should choose appropriate models or develop new models adapted to specific circumstances, considering "the most relevant aspect of reality in a given context" (2015: 11). The reason for this is that social life is complex and contingent (2015: 67, 116): "there are few immutable truths in economics" (2015: 148).

Additionally, I repeat, I have suggested that economic theory should deal with ends. Ends or final causes are needed in order to have actions, to trigger efficient or "pushing" causes or means. We need to know them in order to fully understand economic phenomena. New reverse imperialist economic currents are actually doing this, implicitly recognizing that if we want to carefully explain and try to predict economic phenomena, we need to take into account all motivations of economic actions, thus opening the black box of preferences.

3. The aim of normative economics is *to define the desirable ends of economic actions, to prescribe* ends, including ethical considerations. It should be tied tightly to ethics. In my view, it is a transdisciplinary field containing economic and ethical insights.

   I have argued (2019) that we should consider ethical values as natural facts of the human and social realms, thus legitimizing their presence in social sciences, including economic sciences. This perspective supports Hilary Putnam's position regarding the descriptive and evaluative dual role of ethically "thick" concepts, that is, concepts that involve both descriptive and ethical components.[2] A standard example is "cruel." I maintain that whenever there is a connection between theory and a "thick" ethical concept, ethics

[2] See Bernard Williams ([1985] 1993: 129, 140, 143–145).

must be present. However, not all propositions and concepts of social sciences contain ethical values. As Philippe Mongin (2006: 274) states, "not everything in economics can be evaluative, simply because not every predicate of economics is evaluative."

4. Finally, the aim of practical economic sciences is *to achieve the defined ends*, taking into account all means that may produce the desired effects and, consequently, all motives influencing economic phenomena. I have also upheld that it is a transdisciplinary approach, with a specific method. It integrates elements of economic theory, calculation skills, empirical information, past people's responses to economic and political measures, inside information, etc. Esther Duflo's "economist as plumber" metaphor seems to truly apply to this science:

> Many of us chose economics because, ultimately, we thought science could be leveraged to make a positive change in the world. There are many different paths to get there. Scientists design general frames, engineers turn them into relevant machinery, and plumbers finally make them work in a complicated, messy policy environment. As a discipline, we are sometimes a little overwhelmed by "physics envy", searching for the ultimate scientific answer to all questions – and this will lead us to question the legitimacy of plumbing. This essay is an attempt to argue that plumbing should be an inherent part of our profession: we are well prepared for it, reasonably good at it, and it is how we make ourselves useful. A feature unique to economics is that scientists, engineers, and plumbers all talk to each other (and in fact are often talking to themselves – the same economist wearing different hats). This conversation should continue: it is what will keep us relevant and, possibly, honest. (2017: 31)

Huei-chun Su and David Colander have criticized Duflo's economist as plumber metaphor and have suggested that a better and broader metaphor for Duflo's purpose would be the economist as a "general contractor" who is in contact with theorists, engineers, plumbers, etc., and "she also knows, and works with, the customer of the policy, integrating his values into the design of the policy" (2021: 309).[3]

[3] Philippe van Basshuysen (2021) considers the roles of the designer, the engineering, and the plumber in building institutions.

The four disciplines deal with economic phenomena in different ways. All of them address economics in a broad sense, but a part of theoretical science, positive economics, specifically deals with the economy in the restricted sense. In my 2020 book I conceived these disciplines as sharing specific disciplinary relations.

Having described these notions, let us proceed now to Aristotle, Cartwright, Lawson, and Mäki's.

Aristotle on Economic Science[4]

In Chapter 4, I explained that Aristotle considers economic science as a science in an analogical sense of *oikonomiké*, and that it is a practical science in its specific Aristotelian meaning. At the very beginning of the *Nicomachean Ethics*, he states (I, 2, 1094b 4–6) that Politics:

> ordains which of the sciences should be studied in a state (…) and we see even the most highly esteemed of capacities fall under this, e.g., strategy, economics, rhetoric; now since politics uses the rest of the sciences, and since, again, it legislates as to what we are to do and what we are to abstain from, the end of this science must include those of the others, so that this end must be the good for man.

As it becomes clear from this paragraph, for Aristotle, politics is the "most architectonical" (*Nicomachean Ethics* I, 2, 1094a 27) Aristotelian practical science.

Practical science, as also explained in Chapter 4, shares the characteristic common to all forms of sciences, i.e., to be a "state of capacity to demonstrate (*héxis apodeiktiké*)," (*Nicomachean Ethics* VI, 3, 1139b 32) with the limitations inherent to its subject matter, human choice and human action (contingent, variable, free, singular). Aristotle recognizes this "weaker" character. He asserts in *Nicomachean Ethics*:

> Our treatment discussion will be adequate if it has as much clearness as the subject-matter admits of; for precision is not to be sought for alike in all discussions, any more than in all the products of the crafts. Now fine and just actions, which political science investigates, exhibit much variety and fluctuation (…). We must be content, then, in speaking of

[4] I draw some material from Crespo (2006).

> such subjects and with such premises to indicate the truth roughly and in outline. (*Nicomachean Ethics* I, 3, 1094b 11–27)

Aristotle identifies two reasons for this "inexactness" of practical sciences: "variety and fluctuation" of actions. That is, there are multiple possible situations and human beings may change their decisions. As a result, Aristotle views human action as always singular. He also states:

> We must, however, not only make this general statement, but also apply it to the individual facts. For among statements about conduct those which are general apply more widely, but those which are particular are more true, since conduct has to do with individual cases, and our statements must harmonize with the facts in these cases. (*Nicomachean Ethics*, II, 7, 1107a 31–33)

That is, Aristotle indicates that there is a trade-off between truth and generality. He also affirms that "actions are in the class of particulars, and the particular acts here are voluntary. What sort of things are to be chosen, and in return for what, it is not easy to state; for there are many differences in the particular cases" (*Nicomachean Ethics*, III, 1, 1110b 6–8).

Actions may possess varying properties. An action may be just or unjust according to the situation; and the concrete determination or content of a just situation is also variable (cf., e.g., *Nicomachean Ethics*, V 10, 1137b 28–30 on equity: "... about some things it is impossible to lay down a law ... For when the thing is indefinite the rule is also indefinite"). This is why he says, for example, that "a young man is not a proper hearer of lectures on political science; for he is inexperienced in the actions that occur in life, but its discussions start from these and are about these" (*Nicomachean Ethics* I, 3, 1095a 2–4). He often compares Politics with medicine in this respect, as in the next quotation). In sum,

> matters concerned with conduct and questions about what is good for us have no fixity, any more than matters of health. The general account being of this nature, the account of particular cases is yet more lacking in exactness; for they do not fall under any art or set of precept, but the agents themselves must in each case consider what is appropriate to the occasion, as happens also in the art of medicine or of navigation. (*Nicomachean Ethics* II, 2, 1104a 4–9)

The more "practical" practical sciences are, the less general they become. By leaving generality behind to move toward concrete reality, science becomes narrower in scope. This must be borne in mind; we should adopt a more balanced position: if we try to include all relevant factors in a concrete case, we run the risk of losing generality and, thus, explanatory power upon reaching certain conclusions. But as we try to gain generality, we lose contact with actual reality, and thus explanatory, predictive, and normative "efficiency." Moreover, can we speak of prediction in the above-described conditions? How can this choice between accuracy and generality be solved?

There are two sources that allow generalization in practical sciences: first, an anthropological natural basis; second (and compatible with the former), the recurrence of habits. This is because in the realm of human action "in most respects the future will be like the past has been" (*Rhetoric* II 20 1394a 7–8). Hence, generalizations in practical science are actual dispositions or habits. This accounts for the need for closeness to facts. The more stable the habits and tendencies, the more predictable the outcomes. In any case, general tendencies may change: they are not firmly established universals. Aristotle develops a theory about the stability of habits (*Nicomachean Ethics*, VII, 9, 1151b 25–27 and VII, 10, 1152 a, 26–27). When habits are sufficiently stable so as to constitute social institutions, practical science is firmly based. Therefore, institutions are very important for they consolidate tendencies and habits and facilitate accurate science.

In Chapter 4, I held that practical science occupies a middle ground between prudence (a habit) and theoretical science. In effect, while prudence deals with the particular, practical science tries to formulate generalizations. Despite the fact that generalizations cannot be exact but only probable, practical science should make it its task to elaborate them as valuable resources to tackle its subject matter. As Richard Kraut asserts Aristotle "is asking us to have different expectations of different fields: not *higher* standards for some fields and *lower* for others, but *different* standards" (2006: 87). No matter how different, I insist, the role of practical sciences is to attempt to formulate generalizations. Politics, Aristotle states, "legislates as to what we are to do and what we are to abstain from" (*Nicomachean Ethics* I, 2, 1094b 4–6). Ideally, Reeve comments, "the scope of deliberation should be minimal; that of universal law maximal" (2006: 211). However, the tension remains: inasmuch as practical science tends to universality, it moves further away reality.

Other characteristics of practical science are the following. Firstly, practical science must be closely linked to concrete cases. "Now no doubt," Aristotle says, "it is proper to start from the known. But 'the known' has two meanings—'what is known to us,' which is one thing, and 'what is knowable in itself,' which is another. Perhaps, then, for us at all events, it is proper to start from what is known to us." (*Nicomachean Ethics* I, 4, 1095b 2–4). That is, we must start from the manifest, surface facts to discover the causes.

Secondly, another distinctive feature of practical sciences is their pragmatic end. Aristotle states that "the end of this kind of study [Politics] is not knowledge but action." (*Nicomachean Ethics* I, 3, 1095a 6) and that "we are not conducting this inquiry in order to know what virtue is, but in order to become good" (*Nicomachean Ethics* II, 2 1103b 27–28). He adds in his *Metaphysics* that "the end of theoretical knowledge is truth, while that of practical knowledge is action" (II, 1, 993b 21–22). Nowadays, social sciences are theoretical studies of practical subjects. Hence, we can ask: what is their epistemological basis? Aquinas contributes to Aristotle on this point: he distinguishes three principles to determine whether a science is theoretical or practical: the subject matter, the end, and the method. This threefold classification leaves room for "mixed" cases, such as the theoretical studies of practical subjects just mentioned above. This is an important point because current social sciences—including economics—albeit seeking primarily a theoretical aim, are virtually ordered toward action. Thus, although a particular science may be theoretical *secundum finem*, or may have both theoretical and practical aspects, its implicit orientation toward action determines its epistemological framework.

The third characteristic of practical sciences is normativity. Inexactness, closeness to reality and pragmatic aim are features of practical science stemming from the singularity of human action, as conceived by Aristotle. Besides, the normative character of practical science is associated with its pragmatic purpose. The statement that "it is rational to act in a concrete way" is both a "positive" and normative statement. For Aristotle, practical sciences have a strong ethical engagement.

Finally, a reference should be made to the methodological devices proper to practical sciences. The abundant bibliography on this topic could be summarized in a proposal of methodological plurality. In his *Politics* and *Nicomachean Ethics*, Aristotle admirably combines axiomatic deduction, inductive inference, dialectic arguments, rhetoric, imagination, examples, and topics. Some authors, led by Burnet (1900), have stated that predominance of dialectics and topics stripped practical science of any scientific character (see e.g., Hennis 1963). Aristotelian scholars responded with an in-depth study of the method of practical science. This study yielded the following results: (a) *Politics* and *Nicomachean Ethics* include typical scientific demonstrative arguments (see, e.g., Barnes 1980, pp. 494–495): they are "a state of capacity to demonstrate" (*Nicomachean Ethics* VI, 3, 1139b 31); (b) practical science offers, as any other science, *pisteue*, that is certainty (*Nicomachean Ethics* VI, 3, 1139b 34); (c) it can be transmitted, as all other sciences: this is why Aristotle writes *Politics* and the *Nicomachean Ethics*; (d) it features a hierarchical order: all practical sciences are subordinated to politics (*Nicomachean Ethics* I, 2, passim and *Politics* III, 12, 1282b 14) and they are (implicitly) subordinated to metaphysics, "the most authoritative of the sciences" (*Metaphysics* I, 2, 982b 5).

Summing up, and in view of all previous considerations, we may affirm that, for Aristotle, economic science, as a practical science, is an inexact science that takes into account ethical aspects, it is truer when dealing with locally or temporally limited subjects, and it is methodologically plural. The economic sciences described in the previous section are somehow present in Aristotle's conception of the economy and economic science. Being a practical science, data are relevant to contact with reality. For the Aristotelian practical science, discerning actual ends is central. In addition, practical science is normative and ethically committed. Finally, the pragmatic end of practical science evokes the art of political economy. To conclude, this conception of science fits into the typology of economic science that I proposed in my 2020 book.

Cartwright on Economic Science

In *Nature Capacities*, Cartwright sustains that both the natural and social sciences belong to a world that is governed by capacities which cannot be made sense of without them (1989: 2). However, it seems that there are some differences between the two types of sciences.

Cartwright introduces the notion of a "nomological machine" that proves essential to her conception of science. In *The Dappled World* she defines it in this way:

> It is a fixed (enough) arrangement of components, or factors, with stable (enough) capacities that in the right sort of stable (enough) environment will, with repeated operation, give rise to the kind of regular behavior that we represent in our scientific laws. (1999: 50)

We would need nomological machines to postulate economic laws. However, nomological machines might fail for three possible reasons: (1) lack of fixed enough arrangement of its components, (2) lack of stability in capacities, and (3) lack of stability in environment or circumstances (1999: 49). Capacities, their combination and circumstances are more prone to change in the social field. Consequently, generalizations in social sciences are highly improbable.

Another relevant notion for Cartwright is interaction, "when the tendencies associated with a given factor are changed in the presence of another" (Cartwright 1995: 180). She considers this case in *Nature's Capacities*: "the property that carries the capacity interacts with some specific feature of the new situation, and the nature of the capacity is changed" (1989: 163). Cartwright uses John Maynard Keynes ideas about a "holistic" world in order to support her skepticism about social capacities for this reason (Cartwright 1989: 156–158). Keynes repeatedly refers to "the principle of organic unity," that is to say, the idea that a state of affairs as a whole cannot be analyzed into parts. He affirms, as already quoted in Chapter 8:

> The atomic hypothesis which has worked so splendidly in physics breaks down in psychics. We are faced at every turn with the problems of organic unity, of discreteness, of discontinuity – the whole is not equal to the sum of the parts, comparisons of quantity fail us, small changes produce large effects, the assumptions of a uniform and homogeneous continuum are not satisfied. Thus the results of Mathematical Psychics turn out to be derivative, not fundamental, indexes, not measurements, first approximations at the best. ([1926] 1972: 262)

Circumstances leading to particular combinations affect the stability of causes. Cartwright claims that "most of what happens in the economy is a consequence of the interaction of large number of factors" (Cartwright 2001: 279).

Chapter 4 anticipated some thoughts of Cartwright about economic science. She thinks that economics does not possess basic capacities or principles. What do economists say when confronted with this problem?

Economists are well aware of the need for stable causes. Cartwright presents the example of the Cowles Commission's vision of econometrics: "Econometrics arises in an economic tradition that assumes that economic theory studies the relations between causes and effects" (1989: 149). Econometricians also assume that these causes are stable, like Cartwright's capacities, always acting though not always observable (1989: 150ff.). They also assume that those causal relations are autonomous, i.e., they do not depend on other relations (1989: 155). Cartwright rejects these assumptions.

In addition, according to Cartwright, economists build "over-constrained" models (2007a: Essay V, especially 73–74, 2007b: Chapter 15, especially 219, and 2009: 48–50) that are too "simple" or "sparse," not "simplified" representations of reality (2007a: 70, 2009: 46), in the sense that they are not Galilean idealizations. Galilean idealizations are abstract theories that strip away disturbing causes to look for a key causal factor. Unlike physics, Cartwright notes, economics has very few uncontroversial principles or basic—not derived—capacities at its disposal. In economic models, we thus use only a few principles (usually, the maximizing principle). Then, given that paucity of economic principles with serious empirical content, economic models need to make many unrealistic assumptions "in just the wrong way" (2007a: 78, 2009: 57). Why is this way wrong? We cannot build a model with the maximizing principle as the only constraint. We need to postulate several assumptions. Thus, she asserts, "we can read out only special-case conclusions, not general claims about the manifest results of the capacity" (2007a: 75, 2009: 50). Consequently, "the results of the model are over-constrained [and] (...) the manifest results depend intimately on 'extraneous' factors—factors beyond those that define a Galilean experiment" (2007a: 74, 2009: 49). As a result, the conclusions of economic models are not applicable to real situations (2007a: 78 and V passim and 2007a: 57): "the unrealistic structural assumptions of the model are intensely relevant to the conclusion. Any inductive leap to a real situation seems a bad bet" (2007a: 70, 2009: 45). The models "buy internal validity [rigor: 2007b: 234–235] at the cost of external validity" (2007b: 221).

Things, however, are never black and white. Cartwright helps us arrive at a solution. She states:

> Social science is hard, but not impossible. Nor should that be surprising; natural science is exceedingly hard and it does not confront so many problems as social science — problems of complexity, of reflexivity, of lack of control. Moreover the natural sciences more or less choose the problems they will solve but the social sciences are asked to solve the problems that policy throws up. (2007b: 42)

Although practicing social science is harder than practicing natural science, it is not impossible. According to Cartwright, as mentioned, given the paucity of economic principles, economists move in the wrong direction by adding assumptions instead of looking at the structural circumstances that give rise to particular economic interactions. Economics should consequently consider these structural circumstances: social, institutional, legal, and psychological. In sum, economics should be built on a structural ground, dependent on circumstantial conditions.

Lawson on Economic Science

In keeping with his conception of the economy, after pointing out the problems of modern economics—specially the "fallacy of misplaced universality"—Lawson contends that "economics, appropriately oriented and (so) more reasonably prosecuted, is at most a division of a wider social science, not a separate or autonomous science in its own right" (2003: xxiv). Thus, the title of his book, *Reorienting economics*. In his view, the problems of modern economics are associated with an incorrect underlying ontology. Modern economics conceives reality as a closed system of regularities, presupposing atomism and isolationism, while dealing with a highly internally related, intrinsically dynamic, and structured subject matter, that is, the social domain. Modern economics tries to reductively apply mathematical and deductive logic to regularities in order to make predictions.

Instead, Lawson believes that the primary goal of science is to seek out the underlying causes of phenomena: structures, powers, mechanisms as the ways of acting of things, and tendencies of such mechanisms (2003: 144). Social science also depends on their being underlying social structures to uncover. Lawson maintains that human agency and social structures are interdependent, a fact that makes social science necessarily geographical as well as historical (2003: 149). Critical realism proposes a retroductive mode of inference which requires moving from the concrete phenomena to its mechanisms and underlying structures.

Concerning economics, Lawson specifically considers Mill, Marshall, and Robbins's definitions and proposes a synthesis of them. He states (2003: 154):

> Economics is the identification and study of factors, and in particular social relations, governing those aspects of human action most connected to the production, distribution and use of the material conditions of well-being, along with the assessment of alternative possible scenarios.

It would seem that he considers the study of the economy both in the broad and the narrow sense, as defined in Chapter 4. However, he also asserts (2003: 162, cursives in the text):

> On this conception economics must take its place as a *branch* of social science [...] Its *raison d'être* is not a separate *domain* of distinct phenomena with their own properties, but a particular aspect of *all* social life.

For the reasons I have argued in Chapter 4, I cannot but disagree: Lawson seems to be venturing into the economic imperialist ethos he presumably rejects. However, I cannot but agree with his idea that economics is based on structures and powers, and with its historical-geographical character.

Mäki on Economic Science

In Chapter 4, I explained that Mäki holds that scientific economics deals with economic *commonsensibles*, that is, with "entities that are very much part of the stuff of our commonsense experience" (1998: 306): business firms and households, goods and their prices, money, markets, investments, unemployment, taxes, exchanges, inflation... For Mäki, *Commonsensibles* are not necessarily observable. They "are more broadly characterized in terms of common-sense frameworks and experiences," thus including, for example, preferences and expectations (2000: 111).[5]

[5] Kevin Hoover (1995) has challenged the continuity between commonsensibles of folk economics and scientific economics in relation to macroeconomic entities, while Francesco Guala (2012) and D. Wade Hands (2012) have followed suit in regard to preferences.

"The ontological furniture of 'folk economics' is shared by 'scientific economics'" (1998: 307). Mäki suggests three levels of thought:

> First, there are *general folk views* of "man" and society at large. They comprise folk psychology, among other elements. These folk views are widely and flexibly applicable. Second, there is what may be called *folk economics*. It is a modification and application of the general folk views to economic matters. Third, there is *scientific economics*. Scientific economics accomplishes a modification and rearrangement of the ordinary objects of folk economics. (1996: 434)

I would like to note that Mäki is following the path indicated by Aristotle in his *Topics* (II 2 110a 19–22), as explained in Chapter 4: we should first focus on the folk views and only then apply a scientific analysis to the folk objects.

According to Mäki, scientific economics uses the method of isolation, "whereby a set of elements is theoretically removed from the influence of other elements in a given situation" (1992: 318). Economics uses this approach to analyze common sense entities. The same applies to "ontological furniture": "no new ***kinds*** of entities or properties are introduced" (1996: 435). For example, the "economic man" is an isolated version of ordinary humans (1992: 334). Maki says: "an isolated theory or statement is true if it correctly represents the isolated essence of the object," as the only condition for isolation to lead to the path of truth (1992: 344; see also 1998: 311). What should matter is "significant truth, truth about the essentials of the situation" (1998: 312). Consequently, the "way the world works" imposes constraints to theory, including "imperfections" (2001: 378, 383). Specifically, Maki upholds that economic theories should identify the causal processes that are behind economic phenomena. He applies these criteria to economic models (see 2009, 2011).

Hoover points out that macroeconomic variables do not refer to "ordinary" economic entities like prices, employees, or firms; they are instead complex technical constructs. For Guala, preferences cannot be explained only by introspection, thus introducing theoretical unobservable entities and properties. For Hands, the meaning of terms such as preference, utility, and choice in the contemporary revealed preference theory is radically different from their common sense meaning. Davis (2013), instead, supposes the continuity thesis concerning the concept of preferences involved in Pareto judgments.

Conclusions

We may conclude that the nature of economic sciences results from the nature of the economy, which is contingent and linked to historical, geographical, social, cultural, and psychological conditions, that is, context-dependent. The previous chapters have established that the economic realm is composed of accidental entities or properties that cannot be separated both from underlying variable structures, capacities, and institutions, and from agents having diverse motivations that influence economic aspects. Thus, we face a contingent realm that calls for local analyses. Even Mill, who sustains a primitive form of positive economics, recognizes that it is an abstract science recognized as such for scientific reasons and that we need to allow non-economic motivations to re-enter the scene in order to approximate reality.

Aristotle's notion of economic science as a practical science approaches the economic realm from this perspective. Cartwright, on the other hand, stresses the need to consider the underlying, non-specifically economic dimensions upon developing economic science.

Regardless of my expressed disagreement with Lawson's conception of economics, he remarks that we should search for the causes behind phenomena: structures and powers, mechanisms as a way of acting of a structured thing, and tendencies of such mechanisms. He holds that human practice and social structures are interdependent thus, making social science necessarily historical-geographical (2003: 149). Finally, Mäki, who deals with positive economics, emphasizes the need to find the essential causal mechanisms behind economic phenomena.

In light of the preceding analyses, I would not like to end this chapter without emphasizing the relevance of institutions for economic science. First, I highlighted the institutional dependence of economic behavior. Second, institutions decrease the level of contingency in the economic realm, thus reducing uncertainty and facilitating good economic explanations, forecasting, and adequate policies. As Hindriks and Guala have recently proposed, "institutions feature norms that constrain and enable" (2021: 2027). Institutions are solutions to coordination problems with multiple balances, so typical in the economic field (2021: 2029).

My 2020 book's proposal highlights the relevance of closeness to reality by considering statistics and economic history as economic sciences, the need to discover causes in positive economics, the quest for ends of economic actions as part of economic theory, the role of normative economics and the incorporation of the art of political economy as an economic science.

References

Aristotle. (1995). *The Complete Works of Aristotle: The Revised Oxford Translation*, edited by J. Barnes. Princeton University Press, 6th printing with corrections.

Backhouse, R. E. and S. Medema (2009). "Retrospectives: On the Definition of Economics", *Journal of Economic Perspectives* 23/1: 221–233.

Barnes, J. (1980). "Aristotle and the Methods of Ethics", *Revue Internationale de Philosophie* 133–134: 490–511.

Boumans, M. J. (2015). *Science Outside the Laboratory*. Oxford: Oxford University Press.

Burnet, J. (1900). *The Ethics of Aristotle*. London: Methuen.

Cartwright, N. (1989). *Nature's Capacities and their Measurement*. Oxford University Press.

Cartwright, N. (1995). "Reply to Eells, Humphreys and Morrison," *Philosophy and Phenomenological Research* 55/1: 177–181.

Cartwright, N. (1999). *The Dappled World. A Study of the Boundaries of Science*. Cambridge University Press.

Cartwright, N. (2001). "Ceteris paribus laws and socio-economic machines". In: Uskali Mäki (ed.) *The Economic World View. Studies in the Ontology of Economics*. Cambridge University Press, pp. 275–292.

Cartwright, N. (2002). "The limits of causal order, from economics to physics". In: Uskali Mäki (ed.) *Fact and Fiction in Economics. Models, Realism and Social Construction*. Cambridge University Press, pp. 137–151.

Cartwright, N. (2007a). *Causal Powers: What Are They? Why Do We Need Them? What Can and Cannot be Done with Them?*, Contingency and Dissent in Science Series. London: Centre for Philosophy of Natural and Social Science, LSE, on line in http://www.lse.ac.uk/collections/CPNSS/projects/ContingencyDissentInScience/DP/CausalPowersMonographCartwrightPrint%20Numbers%20Corrected.pdf.

Cartwright, N. (2007b). *Hunting Causes and Using Them. Approaches in Philosophy and Economics*. Cambridge University Press.

Cartwright, N. (2009). "If No Capacities Then No Credible Worlds. But Can Models Reveal Capacities?", *Erkenntnis* 70: 45–58.

Coase, R. (1978). "Economics and Contiguous Disciplines", *The Journal of Legal Studies* 7/2: 201–211.

Crespo, R. F. (2006). "The Ontology of 'The Economic': An Aristotelian Analysis", *Cambridge Journal of Economics* 30/5: 767–781.

Crespo, R. F. (2013). *Philosophy of the Economy. An Aristotelian Approach.* Dordrecht: Springer.

Crespo, R. F. (2019). "Liberal Naturalism and Non-Epistemic Values", *Foundations of Science* 24/2: 247–273.

Crespo, R. F. (2020). *The Nature and Method of Economic Sciences. Evidence, Causality, and Ends.* Routledge.

Davis, J. B. (2013). "Economists' Odd Stand on the Positive-Normative Distinction: A Behavioral Economics View". In: G. DeMartino and D. McCloskey (eds.) *Handbook on Professional Economic Ethics: View from the Economics Profession and Beyond.* Oxford: Oxford University Press, pp. 200–218.

Duflo, E. (2017). "The Economist as Plumber", *NBER Working Paper Series*, Working Paper 23213. https://www.nber.org/papers/w23213. Accessed 12 May 2019. Published in the *American Economic Review* 107/5: 1–26.

Elgin, C. Z. (1996). *Considered Judgment.* Princeton: Princeton University Press.

Fraser, L. M. (1937). *Economic Thought and Language.* London: Adam and Charles Black.

Guala, F. (2012). "Are Preferences for Real? Choice Theory, Folk Psychology, and the Hard Case for Commonsensible Realism". In: Aki Lehtinen, Jaakko Kuorikoski and Petri Ylikoski (eds.) *Economics for Real: Uskali Maki and the Place of Truth in Economics.* Abingdon: Routledge, pp. 137–155.

Hands, D. W. (2012). "Realism, Commonsensibles, and Economics: The Case of Contemporary Revealed Preference Theory". In: Aki Lehtinen, Jaakko Kuorikoski and Petri Ylikoski (eds.) *Economics for Real: Uskali Maki and the Place of Truth in Economics.* Abingdon: Routledge, pp. 156–178.

Hennis, W. (1963). *Politik und praktische Philosophie. Eine Studie zur Rekonstruktion der politischen Wissenschaft.* Neuwied am Rhein and Berlin: Hermann Luchterhand Verlag.

Hindriks, F. and F. Guala (2021). "The Functions of Institutions: Etiology and Teleology", *Synthese* 198: 2027–2043.

Hoover, K. (1995). "Is Macroeconomics for Real?", *The Monist* 78: 235–257; reprinted in U. Mäki (ed.) *The Economic Worldview.* Cambridge: Cambridge University Press, 2001, pp. 225–245.

Keynes, J. M. ([1926] 1972). "Francis Ysidro Edgeworth", *The Economic Journal*, in Keynes, J. M. (1972) *Essays on Biography, The Collected Writings of John Maynard Keynes*, Volume X. London: MacMillan, pp. 251–266.

Keynes, J. M. (1973). *The General Theory and After: Part II. Defence and Development, The Collected Writings of John Maynard Keynes*, vol. 14. London: Macmillan.

Keynes, J. N. ([1890] 1955). *The Scope and Method of Political Economy*, fourth Edition. New York: Kelley and Millman.

Kirzner, I. M. (1976). *The Economic Point of View. An Essay in the History of Economic Thought*, 2nd edition. Kansas City: Sheed and Ward.

Kraut, R. (2006). "How to Justify Ethical Propositions: Aristotle's Method". In: R. Kraut (ed.) *The Blackwell Guide to Aristotle's Nicomachean Ethics*. Oxford: Blackwell Publishing, pp. 76–95.

Lawson, T. (2003). *Reorienting Economics*. London: Routledge.

Mäki, U. (1992). "On the Method of Isolation in Economics", *Poznam Studies in the Philosophy of Science and Humanities* 26: 317–351.

Mäki, U. (1996). "Scientific Realism and Some Peculiarities of Economics". In: R. S. Cohen, R. Hilpinen and Q. Renzong (eds.) *Realism and Anti-Realism in the Philosophy of Science*. Kluwer, pp. 427–447.

Mäki, U. (1998). "Aspects of Realism About Economics", *Theoria* 13/2: 301–309.

Mäki, U. (2000). "Reclaiming Relevant Realism", *Journal of Economic Methodology* 7/1: 109–125.

Mäki, U. (2001). "The Way the World Works (www): Towards an Ontology of Theory Choice". In: U. Mäki (ed.) *The Economic World View: Studies in the Ontology of Economics*, Cambridge: Cambridge University Press, pp. 369–389.

Mäki, U. (2009). "Missing the World. Models as Isolations and Credible Surrogate Systems", *Erknntnis* 70: 29–43.

Mäki, U. (2011). "Models and the Locus of Their Truth", *Synthese* 180: 47–63.

Menger, C. (1889). 'Grundzüge einer Klassifikation der Wirtschaftswissenschaften' (*Jahrbücher für Nationalökonomie und Statistik*, ed. J. Conrad, New Series, Jena: Gustav Fisher, XIX: 465–496.

Menger, C. (1960). 'Toward a Systematic Classification of Economic Sciences'. In: Louise Sommer (transl. and ed.) *Essays in European Economic Thought*. Princeton, NJ: D. van Nostrand Co, Inc., pp. 1–38.

Mill, J. S. (1882). *A System of Logic, Ratiocinative and Inductive*, eighth Edition. New York: Harper & Brothers.

Mill, J. S. ([1844] 2006). *Essays on Some Unsettled Questions of Political Economy* (Essay V: 'On the Definition of Political Economy; and on the Method of Investigation Proper to It'). In *Collected Works of John Stuart Mill*, Volume 4. Indianapolis: Liberty Fund.

Mongin, P. (2006). "Value Judgments and Value Neutrality in Economics", *Economica* 73: 257–286.

Phelps Brown, E. H. (1972). 'The Underdevelopment of Economics', *The Economic Journal*, 82/325: 1–10.

Polanyi, K. (1971). 'Aristotle Discovers Economy'. In: George Dalton (ed.) *Essays of Karl Polanyi*. Boston: Beacon Press.

Reeve, C. D. C. (2006). "Aristotle on the Virtues of Thought". In: Richard Kraut (ed.) *The Blackwell Guide to Aristotle's Nicomachean Ethics*. Oxford: Blackwell Publishing, pp. 198–217.

Robbins, L. C. (1935). *Essay on the Nature and Significance of Economic Science*, second edition. London: MacMillan.

Rodrik, D. (2015). *Economic Rules*. Oxford: Oxford University Press.
Su, H.-ch. and D. Colander (2021). "The Economist as Scientist, Engineer, or Plumber?", *Journal of the History of Economic Thought* 43/2: 297–312.
Tohmé, F. and R. F. Crespo (2013). "Abduction in Economics", *Synthese* 190/18: 4215–4237.
van Basshuysen, P. (2021). "How to Build an Institution", *Philosophy of the Social Sciences* 51/2: 215–238.
Williams, B. ([1985] 1993). *Ethics and the Limits of Philosophy*. London: Fontana Press.

CHAPTER 11

Conclusions

The aim of this book has been to clarify the nature and scope of economics by determining the nature of its subject matter—the economy. I had previously worked on this enterprise in other occasions, especially in my books published in 2013 and 2020. The second chapters in both books deal with the nature of the economy in general, defining it as an analogical concept with two central notions. A broad notion refers to a field of human affairs dealing with decisions and actions aimed at the satisfaction of human needs with resources—goods and services—whose provision implies a cost. A more focused approach considers doing that in the most efficient possible way. Economic activities can be performed economically or not. The restricted notion is a "refined" subset of the broad concept. The broad notion describes the economy as it is, comprising, as we can see in ordinary life, economically and non-economically executed actions. The restricted notion describes a subset of these actions, but it is primarily normative: it indicates the best way of performing economic actions.

Having established empirically testable criteria to define the economical form of economic actions, we can, as positive economics does, precisely determine the best way of executing an economic action, given the resources and preferences at hand. However, as explained, this is a normative definition that is not always implemented. A plethora of motivations that are not comprised in the set of established criteria for the economical

R. F. Crespo, *The Nature of the Economy*,
https://doi.org/10.1007/978-3-031-02453-5_11

come into play, blurring the precision of positive economics. Economics loses its exactness and becomes a highly context-dependent undertaking. As in logics, you need to restrict the field to recover precision. This has been the conclusion of the mentioned previous books, which has been confirmed by the analysis performed in this book.

My book published in 2020, relying on the ideas posited by John Stuart Mill, Menger, and Neville Keynes, proposed to consider four different economic disciplines performing different tasks. The first task is to provide the empirical stuff required by economic theory from statistics and history of the economy. The second task concerns economic analysis or economic theory—an explanatory endeavor that often tries to be predictive as well. It is performed by positive economics as it is known today, but I suggested adding an analysis of the actual ends of economic actions to obtain a complete explanation of them, given that, as explained above, some ends are not usually considered by positive economics. The third task is normative: dealing with the ends that an individual or an economic policy should pursue. Finally, the art of political economy or applied economics unfolds at another level, taking into account the previous inputs to achieve the intended results. In the previous chapter, I quoted Neville Keynes, who believed that political economy—in its widest sense—includes different inquiries with different methods, depending on the nature of their subjects: "*economic uniformities, economic ideals and economic precepts*" ([1890] 1955: 31, 35). Accordingly, as also quoted in the previous chapter, he proposes this distinction and definitions:

> a *positive science* may be defined as a body of systematized knowledge concerning what is; a *normative or regulative science* as a body of systematized knowledge relating to criteria of what ought to be, and concerned therefore with the ideal as distinguished from the actual; and *art* as a system of rules for the attainment of a given end. ([1890] 1955: 34–35, all Keynes's italics)

The common trait shared by the above-mentioned tasks is inexactness. History is subject to interpretations. Statistical data presuppose an elaboration, a "considered judgment" (Elgin 1996), loaded by theory. Concerning positive economics, I have pointed to the need of restricting the field of research in order to recover some exactness, which has been lost because of the context-dependence of economic motivations. For

Dani Rodrik, for example, economists' essential work is to choose appropriate models or to create new models adapted to specific circumstances, capturing "the most relevant aspect of reality in a given context" (2015: 11). The reason for this is the complexity and contingency of social life (2015: 67, 116): "there are few immutable truths in economics" (2015: 148). Normative economics is dependent on the specific ends decided individually or socially. The art of political economy comprises all the vicissitudes of real economy.

All these conclusions are present in my previous works, as well as repeated and reargued in this book from the perspective of a metaphysical analysis. The specific contribution of this book lies in widening the study of the nature of the economy, analyzing the nature of some of its constituents. Indeed, this analysis confirms that economics is and must be inexact to be rigorous. This inexactness is partially overcome by delimiting the scope of particular economic investigations. It is only partially overcome because the richness of reality cannot be fully captured by science, which needs to abstract the core specificities of every situation.

This book has been chiefly philosophical. In fact, the two chapters following the introduction deal with the metaphysical theories and categories that prove necessary for the rest of the book: things or substances, properties or accidents, kinds or essences or natures, forms or structures, capacities, stable causes or powers, mechanisms and tendencies, processes. Chapter 7 is also entirely philosophical, serving as a preamble to ascertaining the nature of macroeconomic entities in Chapter 8. Philosophy is present as well in all the other chapters.

As summarized in the previous chapter, Chapter 4, exploring the economy "in general," truly referred to economic actions. The conclusion was that the economic realm—the part of human reality dealing with the goods that satisfy human needs and desires—consists of accidental entities or properties that cannot be separated from underlying variable structures, capacities, and institutions, as well as the agents that have diverse motivations influencing the economic aspects. We face a contingent realm that calls for local analyses. Chapter 5, focusing on the identity of the economic agent, points out that the economic agent is not only a "maximizer." An ample view of personal identity provides a large room for contingency.

Zeroing in on the economic agent as worker, Chapter 6 highlights the relevance of considering the subjective dimension of work. Current economics, concerned with labor as a factor of production, neglects its most valuable elements almost entirely, and, consequently, it should rethink its treatment of work. The identity of the economic agent—a human person with a plethora of motivations seeking personal fulfillment and a sense of purpose for his life—has an impact on work that economics cannot but seriously consider.

As it tackles macroeconomic entities, Chapter 8 stresses their metaphysical accidental character. The macroeconomic stuff is a system that consists of interactions between individuals and their environment, featuring properties and actions that cannot be reduced to an aggregation of individual behaviors. These new elements have a strong relationship with cultural, social, and political factors and make macroeconomic behavior vastly context- dependent. Finally, Chapter 9 discusses fiduciary money and underscores the contingent character of its value, which largely depends on people's trust in it.

All these characteristics of the subject matter of economics link it closely to the social, political, cultural, institutional, historical, and geographical dimensions where it—micro- and macroeconomy—unfolds. Economic structures are part of the social, institutional, cultural, and political structures and cannot be addressed without factoring in this imbrication. This conclusion stems from the "economic ontologists'" ideas that have been described in these pages. It also derives from the analysis of the economic agent's identity, macroeconomic entities, and money. The contingent character of the subject matter—i.e., the economy—enforces this way of doing economics. It should not be viewed as a limitation but as a requirement of good science, for, as Aristotle sentenced in reference to practical science and as I quoted in the last chapter, "among statements about conduct, those which are general apply more widely, but those which are particular are more true" (*Nicomachean Ethics*, II, 7, 1107a 31–3).

References

Aristotle. (1955). *The Complete Works of Aristotle. The Revised Oxford Translation*, edited by J. Barnes. Princeton University Press, 6th printing with corrections.

Elgin, C. Z. (1996). *Considered Judgment*. Princeton: Princeton University Press.

Rodrik, D. (2015). *Economic Rules*. Oxford: Oxford University Press.

Index

R. F. Crespo, *The Nature of the Economy*,
https://doi.org/10.1007/978-3-031-02453-5

The manufacturer's authorised representative in the EU is Springer Nature Customer Service Centre GmbH, Europaplatz 3, 69115 Heidelberg, Germany. If you have any concerns regarding our products, please contact ProductSafety@springernature.com

Printed and bound by CPI Group (UK) Ltd, Croydon, CR0 4YY

30/09/2026

02225499-0005